Study Guide to Accompany

Elements of Materials Science and Engineering

SIXTH EDITION

Study Guide to Accompany

Elements of Materials Science and Engineering

SIXTH EDITION

Lawrence H. Van Vlack
University of Michigan
Ann Arbor, Michigan

ADDISON-WESLEY PUBLISHING COMPANY

Reading, Massachusetts • Menlo Park, California • New York
Don Mills, Ontario • Wokingham, England • Amsterdam • Bonn
Sydney • Singapore • Tokyo • Madrid • San Juan

This book is in the Addison-Wesley Series in Metallurgy and Materials Engineering.

Reproduced by Addison-Wesley from camera-ready copy supplied by the author.
Consulting Editors: Morris Cohen and Merton C. Fleming

ISBN-0-201-15308-4
ABCDEFGHIJ-AL-89

To the Student

Engineers expect to solve problems during their professional lives. Whether the problems are mathematical or not, they all involve an initial communication step of obtaining information, followed by its analysis, and then implementing a solution. This Study Guide addresses these steps.

The *Quiz Samples* of the guide let you test your ability to understand available oral or written information about engineering materials. You may check yourself with the *Quiz Checks* in the last section of this Study Guide.

Your text includes Examples of problems that illustrate procedures for solving engineering-type calculations. *Practice Problems* are available at the end of each chapter. Try your hand with them and then check your final answer at the end of the text. If after an honest attempt is made to solve those problems you need help, their *Solutions* are included in this guide. Of course, as a maturing engineer, you must eventually work without instructional support; hence, Test Problems are also included in the text.

Instructors know that there are some concepts, such as crystal structures, phase diagrams, semiconductors, and others that require additional attention or an alternate presentation by some students. Thus, there are selected *Study Sets* included in this guide. Use them if the need arises. They have cleared up initial snags for many previous students.

Please contact me at my University of Michigan address if you have positive suggestions regarding either the text itself, or this Study Guide. Future students will appreciate your input.

Nov. 1988 V^2

STUDY GUIDE

for

ELEMENTS of MATERIALS SCIENCE and ENGINEERING
(6th EDITION)

Contents

QUIZ SAMPLES •

Chapter 1 **INTRODUCTION to MATERIALS SCIENCE and ENGINEERING**

1 A Materials characterized three early stages of civilization. These were ____**____, ____**____, and ____**____ .

1 B ____**____ is considered by historians to be a "democratic material" because of its widespread availability for developing civilizations.

1 C As a generality, one may state that "Engineers adapt ____**____ and/or ____**____ for society's needs."

1 D Metallic elements readily ____**____ valence electrons, while nonmetallic elements accept or ____**____ electrons. Ceramic materials contain ____**____ of metallic and nonmetallic elements.

1 E The elastic modulus for a metal is 100,000 MPa; therefore, it takes an applied stress of ____**____ to introduce 0.1% elastic strain.

1 F Stress is ____**____ per unit ____**____ , while strain is ____**____ per unit ____**____ , and a pascal is a ____**____ per m^2.

1 G The elastic modulus of a rubber band is ____**____-er than that of a copper wire of the same dimensions, because it takes a smaller ____**____ to produce 1% strain.

• Quiz Checks are on pp.195-9.

---

2 A The mass of a chromium atom is ___+ -___ .
(a) 52 g (b) 86 $\times 10^{-24}$ g (c) 52 amu (d) 52 amu/0.6 $\times 10^{24}$
(e) None of the above, rather ______________ .

2 B The charge per electron is ___**___ C; where the unit of ___**___ is equal to one ampere·second.

2 C A molecule has strong ___**___-molecular bonds, and weaker ___**___-molecular bonds.

2 D Among molecules that contain less than a dozen atoms, ___+ -___ has (have) two carbon atoms each.
(a) methanol (b) ethanol (c) formaldehyde (d) ethylene (e) vinyl chloride
(f) None of the above, rather ______________ .

2 E Macromolecules contain thousands of atoms made up of repeating units called ___**___ . A molecule with many such units is called a(n) ___**___ .

2 F Vinyl compounds have the composition of C_2H_3**R**. The **R** of vinyl chloride is ___**___ ; in styrene the **R** is ___**___ ; in ___**___ it is -OH; and in propylene it is ___**___ .

2G During polymerization of vinyl chloride, the addition of each mer destroys *one* double bond and introduces ___**___ single carbon-to-carbon bonds. These numbers are ___**___ double, and ___**___ single bonds for the polymerization of styrene.

2 H A polymer contains many identical repeating units called ___**___ .

2 I Of the following atoms, ___+ -___ may coordinate with other atoms in molecules by only a single covalent bond.
(a) hydrogen (b) sulfur (c) bromine (d) None of these, rather __________ .

2 J An Mn^{2+} ion has a radius, r, of 0.08 nm. It can develop a CN = 6 with anions that have R = ___+ -___ .
(a) 0.20 nm (b) 0.15 nm (c) 0.10 nm (d) None of these, rather __________ .

3 A The unit cells of crystals in the ___+ -___ system(s) have equal dimensions in *two* of the three axial directions.
(a) cubic (b) tetragonal (c) monoclinic (d) hexagonal (e) rhombehedral
(f) None of the above, rather ____________ .

3 B The crystal system(s) not listed in 3A is(are) ___**___ .

3 C The fcc structure(s) that possess(es) a packing factor of 0.74 is(are) ___+ -___ .
(a) metal (b) NaCl (c) diamond (d) CaF_2 of Fig. 3-4.2

3 D MgO has the same structure as NaCl. Therefore, a crystal of MgO has ___**___ times as many Mg^{2+} ions as it has unit cells.

3 E A polymorphic change from hcp to ___**___ will introduce no volume change, while a change from bcc to ___**___ or ___**___ produces a volume contraction.

3 F In a unit cell of a bcc crystal, a point at 0.6, 0.2, 0.45 is identical to a point at ___**___ .

3G Based on the data of Appendix B, the distance from $0,{}^1/_2,0$ to $0,0,{}^1/_2$ in the unit cell of aluminum is ___+ -___ nm.
(a) 0.143 (b) $0.143\sqrt{2}$ (c) 0.143(2) (d) $4(0.143)/\sqrt{2}$ (e) $0.405/\sqrt{2}$
(f) None of the above, rather ____________ .
(g) Impossible to calculate.

3 H The [102] direction that pass through ${}^1/_2,{}^1/_2,{}^1/_2$ also passes through ___+ -___ .
(a) 1,0,2 (b) ${}^1/_2,0,1$ (c) -1,0,-2 (d) 0,0,0
(e) None of the previous points, rather __________ .
(f) Impossible to answer; it doesn't pass through the origin.

3 I The (112) plane that contains location 0,0,1 also contains location(s) ___+ -___ .
(a) 1,0,0 (b) $0,0,{}^1/_2$ (c) $1,0,{}^1/_2$
(d) None of the above, rather______________ .
(e) Impossible to determine since the *z* intercept is not *c*/2 .

3 J The plane that passes through the origin and locations ${}^1/_2,{}^1/_2,{}^1/_2$ and 1,0,1 has Miller indices of ___**___ .

3 K Among the following directions, ___+ -___ is(are) parallel to the $(11\bar{1})$ plane.
(a) $[1\bar{1}0]$ (b) [011] (c) $[8\bar{7}1]$ (d) [112] (e) $[\bar{2}1\bar{1}]$
(f) None of the above, rather ____________ .

3 L Gray tin has the structure of diamond (Fig. 3-2.6a), but with *a* = 0.649 nm. There are ___**___ tin atoms per nm in the [111] direction; but the distance between identical points in the [111] direction is ___**___ nm.

3 M The perpendicular distance between (111) planes in copper is ___**___ nm; this distance is ___**___ nm between adjacent (110) planes and ___**___ nm between adjacent (001) planes.

3 N The x-ray wavelength in Fig 3-8.3(a) is 0.156 nm: therefore, the largest interatomic spacing is ___+-___ nm. (Assume $n = 1$.)
(a) 0.21 nm (b) 0.156/(2 sin 21.9°) (c) 0.156/(2 sin 43.8°) (d) 0.11 nm
(e) None of the above, rather ________ .
(f) Impossible to calculate without data from Appendix B.

4 A Point defects within crystals include ____+-____ .
(a) cation pairs (b) anion vacancies (c) hole-interstitial pairs
(d) split atoms (e) dislocations
(f) None of the above, rather ____________________ .

4 B The slip vector of a screw dislocation is ____**____ to the dislocation line. The adjacent atoms possesses extra ____**____ .

4 C Two-dimensional imperfections include external ____**____ , and boundaries between ____**____ or ____**____ .

4 D The presence of a(n) ____+-____ in a crystal introduces strain energy in the adjacent region because atoms are compressed locally.
(a) interstitial (b) vacancy (c) dislocation (d) ion
(e) None of the above, rather __________ .

4 E A crystal changes to a liquid at the ____**____ temperature; a glass changes to a(n) ____**____ liquid at the glass-transition temperature.

4 F The ____**____ is calculated as the molecular weight divided by the mer weight.

4 G The average molecular weight may be calculated on either a ____**____ basis or on a ____**____ basis; the ____**____ basis gives the higher value.

4 H Doubling the degree of polymerization of PVC increases the mean-square length of the molecule by approximately ____**____ percent.

4 I A solid solution alloy possesses two components but only one phase; likewise, a(n) ____**____ possesses two types of mers within a molecule.

4 J A(n) ____**____ stereoisomer has greater regularity than does a(n) ____**____ stereoisomer.

4 K Most of the vinyls typically possess ____+-____ below their glass-transition temperature.
(a) brittleness (b) linear molecules (c) thermoplasticity (d) crystallinity
(e) None of the above, rather ____________________ .

4 L Vulcanization requires the ____+-____ of polymeric molecules.
(a) branching (b) thermosetting (c) oxidation (d) degradation
(e) None of the above, rather ____________________ .

4 M Several common rubbers have the composition of C_4H_5R. The **R** of butadiene is ____**____ ; in isoprene; it is ____**____ ; and in ____**____ , it is chlorine.
R is $-CH_3$ in ____**____ .

4 N A(n) ___**___ solid is noncrystalline. It is called a ___**___ below its transition temperature and a(n) ___**___ liquid immediately above that temperature.

4 O When compared to a crystalline solid, a glass has excess volume called ___**___ .

4 P Since polymeric molecules unkink under tension, the elastic modulus ___-creases as strain increases.

4 Q For extensive substitutional solid solution, the solute and the solvent atoms must have similar ___+ -___ .

(a) atomic diameters (b) atomic weights (c) electronic properties
(d) atomic numbers (e) None of the above, rather ____________ .

4 R A tin bronze contains 10 w/o Sn; the atomic percent tin will be ___** -er than 10 a/o. An aluminum bronze contains 5 a/o Al; the weight percent will be ___** -er than 5 w/o.

4 S The maximum solubility of carbon in bcc iron is 0.02 w/o. This means one carbon atom per ___**___unit cells of iron.

4 T A ___** -metric compound does not have a fixed ratio of atoms.

5 A A(n) ___**___ is a single phase with two or more components; a mixture is a material with two or more ___**___ .

5 B Two ___**___ curves cross at the eutectic temperature. This provides a ___**___ melting liquid.

5 C The ___+-___ is(are) the line(s)on a phase diagram, above which there is only ___+-___. The ___+-___ is(are) the line(s) at which the solidification starts during cooling.
(a) solid (b) liquid (c) eutectic (d) solidus (e) liquidus
(f) None of the above, rather ______________________ .

5 D Brass is a ______-centered cubic phase of copper plus ___**___ . A solid solution of copper plus tin is called ___**___ .

5 E Sterling silver ___+-___.
(a) is valuable because it is commercially pure silver (b) is face-centered cubic
(c) has a maximum solubility of copper at the eutectic temperature
(d) None of the above.

5 F A binary alloy may have three phases in equilibrium at the ___+-___ temperature and at the ___+-___ temperature.
(a) invariant (b) glass-transition (c) eutectoid

5G Interpolation may be used in a ___**___-phase field to determine the ___+-___ of phases. The phase must have identical ___+-___ .
(a) compositions (b) temperatures (c) amounts
(d) None of the above, rather ______________________ .

5 H The ___+-___ reaction(s) involve(s) two phases above the invariant temperature.
(a) peritectic (b) eutectic (c) eutectoid (d) precipitation
(d) None of the above, rather ______________________ .

5 I The three phases of the eutectoid reaction of the Fe-Fe_3C system are ___**___, ___**___, and ___**___.

6A Alloy(s) of ___+ -___ can contain supersaturated solid solution(s) at room temperature.
(a) 97Cu-3Ni (b) 97Cu-3Zn (c) 97Al-3Cu (d) 97Cu-3Al
(e) 97Zn-3Cu (f) 97Cu-3Sn

6B A 97Ag-3Cu alloy solidifies, but contains some coring. The center of the "core" is richer in ___+ -___ than later solids. A 50Ag-50Cu alloy solidifies with some coring. The first solid to solidify is enriched in ___+ -___.
(a) silver (b) copper (c) neither (d) both (e) no way of knowing

6C To solution-treat a 90Al-10Mg alloy, it is best to set furnace temperature at ___+ -___ °C.
(a) 375°C (b) 515°C (c) 620°C (d) None of the above, rather ____°C.

6D The critical radius for homogeneous nucleation ___** creases___ as the solubility limit is approached from lower temperatures.

6E Heterogeneous nucleation may arise from ___**___, ___**___, and ___**___.

6F The thermal expansion coefficient of a solid generally ___** creases___ as the temperature is raised. Furthermore, the volume expansion coefficient for a solid is generally ___** -er___ than for a liquid.

6G In order to have a linear plot of involving thermal activation as a function of temperature, we must plot ___+ -___.
(a) number, *n*, vs °K (b) log *n* vs. °K (c) *n* vs. 1/°C (d) *n* vs. $°K^{-1}$
(e) none of the above, rather ______________.

6H Diffusivity is the coefficient relating flux to ___**___.

6I Other factors equal, diffusivity is greater when the ___sol- **___ atom is small; the packing factor is ___** -er___; and the melting temperature is ___** -er___.

6J The dimensions of diffusivity are ___**___.

6K In order to have a linear plot of diffusivity as a function of temperature, we must plot ___+ -___.
(a) *D* vs. °K (b) *ln D* vs. °K (c) log *D* vs. °K (d) *D* vs. $°K^{-1}$
(e) None of the above, rather ________________.

7 A In single-phase materials, microstructural variations include grain __**__ , __**__ , and __**__ .

7 B In multiphase materials, microstructural variations include those in **7A**, plus the __**__ and __**__ of the phases.

7 C A larger ASTM grain-size number means __+ -__ .
(a) larger grain diameters (b) more boundary area per unit volume (c) more grains
(d) None of the above. Since grains are not spherical, this grain-size index is empirical.

7 D G.S.#5 has __**__ times as many grains per square inch in a two dimensional section (X100) as G.S.#3. The linear dimensions of the latter are __**__ times as large.

7 E Grain boundaries move toward their centers of __**__ . Therefore, __**__ grains get larger.

7 F The average grain size of a metal __** creases__ rapidly at high temperatures; it __** creases__ at lower temperatures.

7 G Reaction products are commonly located along grain boundaries when __+ -__ . These products are commonly found as dispersed particles within grains when __+ -__ .
(a) diffusion is possible (b) there has been extreme supercooling
(c) the cooling rate is slow (d) None of the above, rather __________ .

7 H In steels, __+ -__ is(are) fcc; __+ -__ is(are) bcc.
(a) α (b) γ (c) $\bar{C}$ (d) δ (e) austenite
(f) ferrite (g) carbide (h) pearlite (i) martensite

7 I The Fe-Fe_3C eutectoid temperature is __**__ °C. The Fe-Fe_3C eutectoid composition is at __**__ % C. The eutectoid reaction during heating is __**__ .

7 J __**__ is a lamellar mixture of __**__ .

7 K A 99Fe-1C alloy may contain proeutectoid __**__ . A 99.5Fe-0.5C alloy may contain proeutectoid __**__ .

7 L The name(s), __+ -__ , is(are) used to label characteristic multiphase microstructure(s) encountered in steel; the name(s), __+ -__ , label(s) individual phase(s).
(a) ferrite (b) pearlite (c) carbon (d)tempered martensite
(e) spheroidite (f) martensite (g) liquid

7 M Typically, extensive supercooling __+ -__ the nucleation rate and __+ -__ the growth rate.
(a) increases (b) has no effect on (c) decreases

7N Diffusion is limited to approximately interatomic distances for ___+-___ to proceed as a solid state reaction.

(a) solution (b) precipitation (c) eutectoid decomposition
(d) None of the above, rather ________________________________ .

7O High-density polyethylene, HDPE, has higher values of ___+-___ than LDPE.

(a) thermal expansion (b) thermal conductivity (c) molecular weights
(d) crystallinity (e) heat resistance (f) modulus of elasticity

8 A Two measures of ductility are ____**____ and ____**____. These are strains that precede ____**____.

8 B The ____**____ strength is the stress required to initiate plastic deformation. Above this stress, the strain is both ____**____ and plastic.

8 C The ____**____ strength is the stress calculated as maximum load per unit of ____**____ area.

8 D In a ductile material the breaking strength, S_b, is less than the ultimate strength, S_u, because ____**____ . . .

8 E ____**____ ratio equals ____**____ strain per unit axial strain. Its maximum value of 0.5 implies constant ____**____ during axial loading.

8 F Shear strain is reported as the ____**____ of the elastic angular displacement.

8G The bulk modulus of elasticity is the ratio of ____**____ to ____**____.

8H Possible slip systems in fcc metals include (111) [*uvw*], where [*uvw*] may be ___+-___.
(a) [111] (b) $[1\bar{1}0]$ (c) $[\bar{1}01]$ (d) [011] (e) [001]
(f) $[01\bar{1}]$ (g) [101]

8 I Metallic fracture is commonly categorized as ____**____ or ____**____ depending on the energy absorbed, and/or the ductility.

8 J Toughness is the integrated product of ____**____ and strain prior to fracture. This gives units of ____**____ per volume.

8 K There is a major ____**____ of toughness for steels below the ____**____ temperature. This characteristic is encountered in ____**____ -centered cubic metals, but not in ____**____-centered cubic metals.

8 L Other things equal, a steel with a ____**____ -grain size will have a lower T_{dt}.

8 M It is necessary to standarize the geometry of test samples for toughness testing, since the required energy depends on the ____**____ and ____**____ of the sample.

8 N For a given nominal tensile stress, the stress-intensity factor increases by a factor of ____+-____ if the crack length is doubled.
(a) two (b) square root of two (c) four

8O A ____**____ is a stress-raiser because it focuses stresses at the ____**____ of the crack. The stress-intensity factor is proportional to the nominal stess and the ____**____.

Ni ↑ yield prop & hardness p290

9 A Specifications call for an ultimate strength of at least 300 MPa and a hardness of no more than 65 R_f. These requirements can be met by __+-__ .

(a) 90Cu-10Zn (b) 90Cu-10Ni (c) 80Cu-20Zn
(d) 80Cu-20Ni (e) 70Cu-30Zn (f) 70Cu-30Ni
(g) None of the above .

9 B Alloy (1) is 99Cu-1Sn (conventional w/o basis). Alloy (2) is 99Cu-1Be. Both are solution treated. The yield strength of alloy (1) is __+-__ that of alloy (2).
(a) the same as (b) less than (c) greater than
(d) not comparable with (because their atomic numbers are different)

9 C Mechanical working includes the processes of __**__ , __**__ , and __**__ .

9 D A 1.1-mm diameter wire is cold drawn through a 1.0-mm die. It is cold worked __+-__.
(a) 10 % (b) 0.1/1.1 (c) 21 % (d) None of those, rather ________ .

9 E A plate of copper must have a hardness ≥125 BHN and a ductility ≥20 % El. (50-mm). Its specification window for cold working is __+-__.
(a) 20-to 30 % (b) 25 to 50 % (c) best at ~40 %
(d) None of these, rather ___________ (e) Impossible by cold working.

9 F Other factors equal, recrystallization occurs in shorter times when the metal __+-__.
(a) has higher purity (b) is strain hardened less
(c) is heated slowly (d) has a high melting temperature

9 G In order to have a linear plot of recrystallization time as a function of temperature, we must plot __+-__.
(a) t vs. K (b) $ln\ t$ vs. K (c) log t vs. K (d) D vs K^{-1}
(e) None of the above, rather ________________ .

9 H __+-__ reactions require the formation of new phases within a solid.
(a) Grain growth (b) Recrystallization (c) Solution
(d) Precipitation (e) Eutectoid (f) Martensitic

9 I To precipitation harden a material, the first step is to __**__ the solubility limit . The second step is to __**__ the solubility limit .

9 J Overaging brings about an increase in the __+-__ precipitates, and a decrease in the __+-__ precipitates.
(a) number of (b) volume fraction of (c) distances between (d) hardness of

9 K Martensite does not form during the steel-treating process called __+-__ .
(a) interrupted quench (b) normalizing (c) full anneal
(d) process anneal (e) quenching

9 L The name(s), __+ -__ , is(are) used to label characteristic multiphase microstructures encountered in steel, while the name(s) __+ -__ label(s) individual phase(s)
(a) ferrite (b) pearlite (c) carbon (d) tempered martensite
(e) spheroidite (f) martensite (g) liquid

9 M Typically, extensive supercooling __+ -__ the nucleation rate and __+ -__ the growth rate.
(a) increases (b) has no effect (c) decreases

9 N Diffusion is limited to approximately interatomic distances for __+ -__ to proceed as a solid state reaction.
(a) solution (b) precipitation (c) eutectoid decomposition
(d) None of the above, rather ____________.

9 O Reaction products are commonly located along grain boundaries when __+ -__. These products are commonly found as dispersed particles within grains when __+ -__.
(a) diffusion is possible (b) there had been extreme supercooling
(c) the cooling rate is slow (e) None of the above, rather ____________ .

9 P When formed in a 99.5Fe-0.5C alloy, martensite is __+ -__ .
(a) stable below the M_f temperature (b)hard and easily fractured
(c) body-centered cubic (d) None of the above.

9Q __+ -__ is(are) mixture(s) of ferrite and carbide found in steels.
(a) pearlite (b) tempered martensite (c) spheroidite (d) None of the above.

9 R Phase changes occur when annealing involves __+ -__ .
(a) glass (b) cold-worked brass (c) stress relief of gears
(d) spheroidization of pearlite (e) None of the above, rather ____________ .

9 S __**__ and __**__ are names of processes used to homogenize metals.

9 T The pearlite layers are __**__ in a normalized steel than in one that is fully annealed, because the cooling rate is more rapid.

9 U High strengths and toughness are difficult to realize simultaneously in steel when __+ -__ is predominant.
(a) spheroidite (b) tempered martensite (c) pearlite (d) martensite

9 V A steel that resists __**__ has high hardness. One that readily attains its maximum hardness has high __**__ .

9 W Steels have greater hardenability, if __**__ are present.

9 X __**__ quenching is more severe than oil quenching, since the former has a very high __**__ .

10A There is a major drop in the viscoelastic modulus at the ____**____ temperature. This temperature is slightly lower when ____**____ time is available.

10B Continued strain over an extended period of time leads to stress ____**____ .

10C The elastic modulus of an oriented polymer is ____**____ than the elastic modulus of an amorphous polymer.

10D For a composite to show improved strength, the reinforcement must be stronger than the matrix. In addition, it must have a higher ____**____ .

10E To a first approximation, the elastic modulus in an oriented composite is proportional to the ____**____ of each component.

10F The properties of ____**____ composites are very directional, or ____**____ .

10G The interface between the reinforcement and the matrix must be well bonded in order to withstand ____**____ stresses.

10H In ____**____ loading, S_u/ρ serves as a basis for design comparisons; in ____**____ loading, and for rigidity, E/ρ is better basis.

10I The glass coating (enamel) on the steel "box" that makes the walls of a kitchen oven should have an thermal expansion coefficient that is ____+ -____ that of the underlying metal.
(a) greater than (b) the same as (c) less than

Chapter 11 — CONDUCTING MATERIALS

11 A A copper wire is to be 1 m long and have a resistance of 2 mΩ. Its conductivity is 60 $\times10^{6}$ $\Omega^{-1}\cdot m^{-1}$. Its cross-sectional area should be __+ -__ .
(a) $\pi(3.25\ mm)^2/4$ (b) $(0.002)(1)/60\ \times10^{6})$
(c) Neither of the above, rather ____________ .
(d) Insufficient data to calculate.

11 B Refer to Problem 8-P22. Before loading, the wire was 3.07 m. Its resistance is __**__ . .With 1.5 V across the wire's length, there will be __**__ electrons entering the wire per millisecond.

11 C A copper wire (1.1 mm dia.) was chosen to meet resistance specifications. How much heavier (+%) or lighter (-%) will a pure aluminum wire be that has the same resistance? __+ -__
(a) -70 % (b) +41 % (c) -48 % (d) -52 %
(e) None of those listed, rather____________ .
(f) Insufficient data to calculate.

11 D The conductivity of copper at 0°C is 63 $\times10^{6}$ $\Omega^{-1}\cdot m^{-1}$; at 100°C is 43 $\times10^{6}$ $\Omega^{-1}\cdot m^{-1}$; and at 200°C (473 K) is __+ -__ $\Omega^{-1}\cdot m^{-1}$. The resistivity of copper is 16 $\times10^{-9}$ $\Omega\cdot m$ at 0 °C; is 23 $\times10^{-9}$ $\Omega\cdot m$ at 100 °C; and is __+ -__ $\Omega\cdot m$ at 200 °C .
(a) 30 $\times10^{-9}$ (b) > 30 $\times10^{-9}$ (c) 23 $\times10^{-9}$ (d) <23 $\times10^{-9}$
(e) None of those listed, rather____________ .
(f) Insufficient data to calculate.

11 E At temperature, *T*, the probability is 0.01 that an energy level at (E_f + 0.1 eV) is occupied. The probablilty of occupancy at (E_f - 0.1 eV) is __+ -__ .
(a) -0.01 (b) +0.01 (c) 1/0.01 (d) 1-0.01 (e) 1+0.01
(f) None of the above, rather ____________ .
(g) Impossible to determine without a specific temperature.

11 F The value of $F(E)$ in a metal decreases with increased temperature at __+ -__ .
(a) E_f + 0.1 eV (b) E_f (c) E_f - 0.1 eV

11 G The mean-free path of electron drift increases with __+ -__ .
(a) purity (b) strain hardening (c) recrystallization (d) elastic modulus
(e) None of the above, rather ____________ .

11 H The energy band of a __**__ is only partially filled, while it is filled in both a __**__ and __**__ .

11 I In a semiconductor, a useful number of electrons can be raised across the energy __**__; very few possess that much energy in a(n) __**__ .

11 J Of two group VI elements that have the same diamond cubic structure, the one with the larger ___+-___ is expected to have the smaller energy gap.
(a) packing factor (b) coordination number (c) number of valence electrons
(d) None of the above, rather ______________ .

11 K The conductivity of a semiconductor is the sum of the conductivity in the ___**___ band by negative carriers, and the conductivity in the ___**___ band by positive carriers.

11 L The simplest compound semiconductors contain elements of Groups ___**___ and ___**___ with an average of ___**___ valence electrons per atom. Those compounds with ___**___ elements have narrower energy gaps.

11 M The numbers of electrons jumping the energy gap depend upon the reciprocal of the ___**___ ___**___ . In order to make a linear plot, the numbers must be expressed in terms of ___**___ .

11 N Electrons can be raised across the energy gap, E_g, by ___**___ , by ___**___ , and by energetic ___**___ . In the first case, the numbers are proportional to $e^{-E_g/2kT}$; in the second, the ___**___- conduction depends upon the number of ___**___ hitting the semiconductor.

11 O When an electron jumps the gap, an electron-___**___ pair is formed. This is eliminated by the ___**___ of the electron and the hole.

11 P The term ___**___ applies to those semiconductors that possess semiconductivites by virtue of impurities; a semiconductor without impurities is called ___**___ .

11 Q A semiconductor device for a ___**___ works on the principle that the energy gap is ___**___ in size with an increase in density, so that more electrons can enter the conduction band.

11 R A ___**___ device depends simple on the fact that more electrons enter the conduction band at higher ___**___ .

11 S Current passes through the junction of a ___**___ when a ___**___ bias is applied.

12A Significant magnetic induction, B, remains in a __**__ magnet after the external magnetic field is removed. A reversed field, called the __**__ , is required to drop the induction to zero.

12B Fe^{2+} and Fe^{3+} ions have six and five 3-*d* electrons, each. Therefore, they have __**__ and __**__ Bohr magnetons, respectively. Vanadium, with three 3-*d* electrons, has __**__ Bohr magnetons.

12C Magnetic domains expand by sideway movements of domain __**__ ; __+-__ interfere with these movements.
(a) dislocations (b) precipitates (c) grain boundaries (d) impurities
(e) None of these, rather ______________________ .

12D To make steel magnetically softer, processing includes __+-__ .
(a) annealing (b) grain growth (c) decarburization (d) quenching
(e) None of these, rather ______________________ .

12E In magnetite, a ferrite found in nature, the iron ions are in __**__ sites, and __**__ sites. The iron ions in the two sites have __**__ magnetic orientation.

12F Since the net magnetic moment of a unit cell of $[NiFe_2O_4]_8$ is ~150 $\times 10^{-24}$ $A \cdot m^2$, there are __**__ unpaired electrons per unit cell.
Since the magnetization of this compound is 260,000 A/m, we can back calculate a = __**__ for the size of the unit cell.

12G The permeability of a magnet is the ratio of __**__ and __**__ . In selecting hard and soft magnets, permeability receives more attention in __**__ .

12H The property of __**__ is the summation of magnetic moments per __**__ . The magnetic moment per electron is __**__ $A \cdot m^2$. For two electrons in the same orbital, the net magnetic moment is __**__ .

12I Ferrites are illustrative of ceramic magnets in which iron atoms are present as __**__ and __**__ . The former has __**__ unpaired electrons; the latter has __**__ unpaired.

12J The term, __**__-magnetic, refers to a material that possesses ions with magnetic moments, but one in which the magnetization is not obvious since there are equal numbers of ions with opposing magnetic orientations.
In a __**__-magnetic material, ions have opposing magnetic orientations, but the magnetic moments are not balanced.

12K A magnetic material possesses __**__ , which are small regions containing unit cells, all with the same magnetic orientations. The __**__ between these are readily moved in a __**__ magnet.

12 L A hard magnet is evaluated on the basis of its ___**___ product, which is a measure of the ___**___ to demagnetize the material.

12 M A cold-worked steel is mechanically hard because of the introduction of ___**___ that become tangled and restrict slip. These steels are magnetically hard because their introduction also restricts ___**___ movements.

12 N Domain boundaries become anchored by ___**___ , by ___**___ , and by ___**___ .

12 O A material loses its magnetization at its ___**___ temperature because ___**___ . . . .

13A Dielectric materials, which are used to separate two __**__ , must have a large __**__ in order to serve as an insulator and not be a semiconductor.

13B Although the units for dielectric strength are__**__, the values for an insulator are generally not proportional to __**__ , because __**__ . . .

13C Dielectric strength can be reduced by __**__ , by __**__ , and by __**__ .

13D Polarization is a measure of __+-__ .
a) product of charge and distance b)voltage gradient to produce electrical breakdown
c) dielectric constant per unit volume
d) None of the above, rather ______________ .

13E The dielectric constant is increased by __**__ , __**__ , and/or __**__ polarization.

13F Several types of polarization may be described; these include __**__ , __**__ , and __**__ . Of these, __**__ can occur at highest frequencies, and __**__ is least rapid.

13G As a property, polarization is the summation of the __**__ per__**__ ; it also is the added __**__ arising from the presence of displaced electrons, displaced ions, and polarized molecules.

13H Relative dielectric constants are always greater than __**__ ; in general, they have __**__ values at higher frequencies.

13I With d.c., the highest dielectric constant for a polymer is obtained just above the __**__ temperature. It decreases at higher temperature because __**__ . . . .
At lower temperatures, __**__ polarization is the principal contributor to the dielectric constant.

13J At high frequencies, __**__ polarization is the principal contributor to the dielectric constant, because __**__ . . .

13K The polarization of a __**__ material can be changed by a stress that introduces an elastic __**__ .

13L The interrelationship between polarization and strain provides a basis for __**__ devices that convert mechanical energy to electrical, and vice versa.

13M For a material to be either __**__ or __**__ , it must lack a center of symmetry so that the centers of positive and negative charges are separated.

13N All __**__-electric materials are __**__-electric, but not all __**__-electric materials are __**__-electric, because __***__ . . . .

13O ___**___ polarization is that amount that remains after any external field is removed; the ___**___ is the reverse field required to drop the polarization to zero.

13P A lower index of refraction accompanies ___+-___ .

a) slower light velocity
b) high electronic polarization
c) greater light absorption
d) None of the above, rather ________________ .

13Q Optical fibers for communication are produced to have a glass with a lower ___**___ at the surface.

14A Failure can arise from ___+-___ as well as normal service wear and tear.
(a) misuse (b) unanticipated service conditions (c) poor quality control
(d) poor design judgements (e) improper maintenance

14B Design consideration(s) include(s) the ___+-___ of the material as well as specific properties required of the material.
(a) cost (b) availability (c) discard, or reuse
(d) combinations of properties (e) environmental implications

14C For corrosion to proceed, there must be a(n) ___**___, a cathode, and a(n) ___**___.

14D Corrosion occurs at the ___**___. The commerical process, called ___+-___, is the corrosion reaction in reverse.

14E Reference electrolytes are ___**___-molar at ___**___ °C. This establishes a standard emf of +0.34 V for copper (Table 14-2.1), but < 0.34 V in ___+-___ electrolytes.
(a) more dilute (b) more concentrated
(c)warmer,0.1M (d) cooler, 0.1 M

14F The most common cathode reaction that leads to rusting of iron in moist air is ___**___.

14G The anode for each of the following pairs is ___+-___. (Unless indicated, concentration temperature, etc., are the same.)
(a) iron vs.(b) copper (c) high O_2 vs. (d) low O_2
(e) dilute solution vs. (f) conc. solution (g) grains vs. (h) grain boundary
(i) annealed vs. (j) cold-worked

14H The oxidation of ___**___ is used as an arbitrary reference for other electrode reactions.

14I Reference electrode solutions have a ___**___ concentration; solutions that are more ___**___ are more anodic.

14J When the electrolyte concentration varies, corrosion occurs more readily where the concentation is more ___**___.

14K In general, oxygen accelerates the rate of corrosion with corrosion occurring most rapidly where the oxygen content is ___**___.

14L On a microscale, the grain boundary becomes the ___**___ and the interior of the grain becomes the ___**___. Likewise, annealed metal is cathodic to ___**___ metal because ___**___ . . . .

14M Electrons move from the ___**___ to the ___**___ through the metallic part of the circuit. Current flows from the ___**___ to the ___**___ through the electrolyte.

14N A surface of a metal becomes ___**___ if it develops a protective oxide film; it becomes ___**___ if the film is destroyed.

14O The combination of ___**___ and electrolyte leads to more rapid corrosion and failure by ___**___ fracture. This type of failure is related to ___**___ fatigue for glass and polymers.

14P Protective surfaces include ___**___ films, as well as coatings of ___**___ , ___**___ , and ___**___ . Also, ___**___ are effective because they supply oxygen-containing ions that are absorbed onto the surface.

14Q Zinc bars bolted onto the sides of ships provide corrosion protection because they become ___**___ . On shore, a dc current can provide protection if connected so that the electrons move ___**___ the metal to be protected (the current flowing in the other direction).

14R Galvanized steel is protected because ___**___ · · · ; anodized aluminum is protected because ___**___ · · · ; stainless steel does not corrode readily because ___**___ · · · .

14S Cyclic stresses can lead to ___**___ if the stress level is above the ___**___ . The test for ultimate strength (Section 8-2) involves ___**___ stress cycle(s).

14T Delayed fracture can occur from ___+-___ even though the service conditions have not been changed.
(a) tension loads alone (b) compression loads alone (c)cyclic loading
(d) creep (e) chemical reactions
(f) None of the above, rather ______________________ .

14U Static fatigue and stress corrosion have ___**___ in common.

14V Endurance limit is a ___+-___ .
(a) stress (b) property of all materials (c) strain (d) fracture mechanism
(e) None of the above, rather ______________________ .

14W Stage ___**___ creep has the most uniform rate of the three stages.
Its creep rate increases with an increase on ___**___ and ___**___ and with a decrease in ___**___ .

14X For oxidation to proceed and for scale to form , there must be ___**___ through the scale. If the scale is adherent and doesn't flake off, the scale `requires ___**___ times as long to become 2 mm thick as it took to become 1 mm thich.

14Y When ___**___ occurs at the surface of steel, the microstructure becomes more ferritic.

14Z During ___**___ , the side atoms of a polymer are lost and only the carbon remains.

14AA Radiation exposure can lead to ___**___ in polumers, by which large molecules are broken into two or more smaller molecules.

14AB Plain-carbon steels that are exposed to neutron radiation develop reduced __+-__ .
(a) ductility transition temperatures (b) yield strength (c) ductility
(d) resistivity (e) None of the above, rather ________________ .

14AC Some widely occurring oxides that serve as refractories include __**__ and __**__ . Less common refractory oxides include __**__ and __**__ .

14AD Spalling resistance is greater for refractories that have high values of __**__ and __**__ , and low values of __**__ and __**__ .

14AE Radiation damage involves the introduction of many __**__ defects. In contrast, __**__ involves the introduction of dislocations, which are linear defects.

14AF Because of the contrast in 14AE, __**__ from radiation damage occurs at lower temperatures than does __**__ after strain hardening.

14AG __**__, caused by radiation, degrades a polymer by reducing the __**__ molecular size.

SOLUTIONS
to
PRACTICE PROBLEMS*

Chapter 1 **MATERIALS SCIENCE and ENGINEERING**

1-P21 What is the stress on a 1-mm (0.04 in.) diameter wire that supports a 3-kg (6.6-lb_f) load?

Procedure : Stress (N/m^2, or Pa) is force per unit area; force is mass x acceleration.

Solutions : $(3 \text{ kg})(9.8 \text{ m/s}^2)/(0.001 \text{ m})^2(\pi/4) = \mathbf{37 \text{ MPa}}$;

or $(6.6 \text{ lb}_f)/(0.04 \text{ in.})^2(\pi/4) = 5{,}300 \text{ psi}$.

1-P22 When the stress on the wire of Problem 1-P21 is 37 MPa (5300 psi), the elastic strain is 0.00054. What is the elastic modulus?

Procedure : The elastic modulus (Young's modulus) is the ratio of stress to strain.

Solution : $E = s/e = 37 \text{ MPa}/0.00054 = \mathbf{69{,}000 \text{ MPa}}$;

or $5{,}300 \text{ psi}/0.00054 = 9{,}800{,}000 \text{ psi}$.

Comment : Since strain is dimensionless, it is independent of units.

1-P23 A wire is 36 ft (11 m) long. It has been stretched 1.38 in. (35 mm). What stress was developed if the elastic modulus is 10,000,000 psi (70 GPa)?

Solutions : $e = 1.38 \text{ in.}/(36 \times 12 \text{ in.}) = 0.0032$

$s = Ee = (10{,}000{,}000 \text{ psi})(0.0032) = \mathbf{32{,}000 \text{ psi}}$;

or $e = 35 \text{ mm}/(11{,}000 \text{ mm}) = 0.0032$

$s = Ee = (70{,}000 \text{ MPa})(0.0032) = 220 \text{ MPa}$.

---

* These *Practice Problems* (1) use equations directly (plug and chug), (2) refer to specific statements in the text, or (3) "mimic" the examples at the end of the sections. You should try them for practice, before attempting the Test Problems in the text, which require the student to work from a fuller understanding.

1-P24 Identify the number of significant figures that you should read from your calculator for Problems 1-P21, 1-P22, and 1-P23.

Explanation : Problems 1-P21 and 1-P22 use diameters with only one sig. fig. (significant figures). Therefore, the results are accurate to 1 sig. fig. only. However, we commonly include one additional digit to indicate which way the figure "tilts," e.g., in Problem 1-P21, >5,000 psi, rather than < 5,000 psi. (Do *not* give 5252.11312 psi, or 37.433243 MPa answers. It would show that you are *not* thinking!!) Problem 1-P23 is somewhat ambiguous, since 70 GPa can be 70 ± 0.5, or 70 ± 5. Use a second digit in this case.

1-P25 Brass has a resistivity of 62 x10^{-9} ohm·m. (a) What is the end-to-end resistance of a brass strip that is 5-cm long by 5-mm wide by 0.5-mm thick? (b) What is its conductivity?

Solution : (a) $R = \rho L/A$ = (ohm·m)(m)/m^2

= (62 x 10^{-3} ohm·m)(0.05 m)/(0.005 m)/(0.0005 m) = **0.0012 ohm** .

(b) $\sigma = 1/\rho$ = 1/(62 x 10^{-9} ohm·m) = **16 x10^6 ohm^{-1}·m^{-1}.**

Comment : Rather than memorizing Eq. (1-2.1), use the dimensional units to set it up .

1-P26 Compare and contrast *strength* , *ductility* , and *toughness* .

Strength:	**stress to failure .**
Ductility:	**strain to fracture .**
Toughness:	**energy for fracture .**

1-P27 Compare and contrast *resistance* , *resistivity* , and *conductivity* . What are the units of each?

Resistance :	ohm	Function of material and shape.
Resistivity :	ohm·m	Function of the material, independent of shape.
Conductivity :	ohm^{-1}·m^{-1}	Reciprocal of resistivity .

1-P41 (a) Cite metal parts that are used in a modern kitchen stove. What property is pertinent to each application? (b) Repeat part (a) for ceramics. (c) Repeat part (a) for polymers.

Metal :	Oven box;	easy shaping by "deep drawing".
	Resistor wire;	selected electrical resistivity.
	Element jacket;	oxidation resistance.
Ceramic :	Insulation between resistance wire and element jacket.	
	Oven box coating;	easy cleaning, oxidation resistance, color.
Polymer :	Dial knobs;	easy shaping, color, cost.
	Power wire coating;	insulation, flexibility.

(5.248 g) (.6 x10^{24} AMU/g) (1 mol / 63.46 g) mol CU

(162 mm $\frac{\pi}{4}$ D^2)

5.248 g (.6 x 10^{-24} AMU/g) (CU / 63.43 g) CU/mm^3

2-P11 (a) What is the mass of a magnesium atom in grams? (b) The density of magnesium is 1.74 Mg/m^3 (1.74 g/cm^3); how many atoms are there per mm^3?

Solutions : (a) (24.3 amu/mag.)/(0.602 x 10^{24} amu/g) = **40.4 x10^{-24} g/magnesium.**

(b) (1.74 g/cm^3)/(40.4 x 10^{-24} g/mag.) = **43 x10^{21} magnesium/cm^3.**

Comment : An amu is a convenient unit in many calculations; however, it is your choice whether you use amu's or grams.

2-P12 A copper wire weighs 5.248 g, is 2.15 mm in diameter, and is 162 mm long. (a) How many atoms are present per mm^3? (b) What is the wire's density.

Solutions : (a) $$\frac{(5.248\ \text{g})(0.602\ \text{x}10^{24}\text{Cu}/63.54\ \text{g})}{(162\ \text{mm})(\pi/4)(2.15\ \text{mm})^2} = \mathbf{84.5\ x10^{18}\ Cu/mm^3}$$

(b) (5.248 g)/[(16.2 cm)(π/4)(0.215 cm)^2] = **8.92 g/cm^3** (or 8.92 Mg/m^3) .

2-P13 (a) How many silver atoms are there per gram? (b) What is the volume of a silver wire containing 10^{21} atoms? (Obtain the density from an appendix.)

Solutions : (a) (0.602 x10^{24}Ag)/107.87 g = **5.58 x10^{21} Ag atoms/g** .

(b) $$\frac{10^{21}\ \text{Ag}/(5.58\ \text{x}10^{21}\text{Ag/g})}{10.5\ \text{g/cm}^3} = \mathbf{0.017\ cm^3}$$ (or 17 mm^3). .

2-P14 Chromium is to be plated onto a steel surface (1610 mm^3) until it is 7.5 μm thick. (a) How much chromium (Cr^{2+}) is required? (b) How many amperes will be required to do this in 5 minutes?

Procedure : Obtain the density from the Appendix. In (b), "walk through" the solution by watching the units.

Solutions : (a) (1610 mm^2)(0.0075 mm)(7.19 mg/mm^3) = **87 mg** (or 0.087 g) .

(b) (0.087 g)(0.602 x10^{24}Cr/ 52.0 g)(2 el/Cr)(0.16 x10^{-18}A·s/el)/(300 s) = **1.07 A.**

2-P15 Distinguish between atomic number and atomic weight.

(a) Atomic number is the number of electrons (and protons) per neutral atom.

(b) Atomic weight is the mass per atom (amu), or per mole (grams).

2-P21 Determine the molecular weight for the molecules of Fig. 2-2.1.

Units : amu per molecule, or g/mole .

Solutions :

(a) CH_3OH	12+4(1) + 16	= **32**	(f) CH_2O	12 + 2(1) + 16	= **30**
(b) C_2H_5OH	2(12) + 5(1) + 17	= **46**	(g) CH_3COCH_3	3(12) + 6(1) + 16	= **58**
(c) NH_3	14 + 3(1)	= **17**	(h) NH_2CONH_2	2(14)+4(1)+16+12	= **60**
(d) C_6H_6	6(12) + 6(1)	= **78**	(i) C_2H_4	2(12) + 4(1)	= **28**
(e) C_6H_5OH	6(12) + 5(1) + 17	= **94**	(j) C_2H_3Cl	2(12) + 3(1) +35.5	= **62.5**

*See the footnote with the *Practice Problems* of Chapter 1.

2-P22 A small diamond has 10^{20} atoms. How many covalent bonds does it have?
Four bonds with each atom; each bond is shared by two atoms. Net = 2 bonds/atom.

10^{20} atoms ∞∞⟹ **2 x10^{20} bonds** .

2-P31 A common polymer has C_2H_3Cl as a mer (Table 2-3.1). It has an average mass of 35,000 amu per molecule. (a) What is its mer mass? (b) What is its degree of polymerization?

Solutions : (a) mer mass = 2(12 amu) + 3(1 amu) + 35.5 amu = **62.5 amu** .
(b) D.P. = *n* = 35,000 amu / 62.5 amu = **560 mer/molecule** .

2-P32 Using the data of Section 2-2, what is the net energy change as 18 g of vinyl alcohol polymerize to polyvinyl alcohol?

Procedure : Change to moles. Data from Table 2-2.2.

Double bonds must be broken (+ energy) 1 C = C broken per mer .

Single bonds formed (- energy); 2 C — C formed per mer .

Solution : 18 g/ (24 + 3 + 17 g/mole) = 0.41 moles
(0.41 moles)(+680 kJ/mole) = +278.2 kJ
2(0.41 moles)(-370 kJ/mole) = -302.7 kJ ; Net = **-24.5 kJ** .

Comment : Energy is required for starting polymerization; but once started, there is net energy release for continued polymerization.

2-P33 There are 9.2 x10^{18} molecules per g of polystyrene. (a) What is the average molecular size? (b) What is the degree of polymerization?

Solutions :
(a) (0.602 x10^{24} molecules/mole)/(9.2 x10^{18} molecules/g) = **65,000 g/mole** ,
or (0.602 x10^{24} amu/g)/(9.2 x 10^{18}molecules/g) = **65,000 amu/molecule** .
(b) (65,400 amu/molecule)/[24 + 3 + (72 + 5) amu/mer] = **630 mers/molecule** .

2-P34 (a) How many C=C bonds are eliminated per mer during polymerization of polyvinyl chloride? (b) How many additional C—C bonds are formed?

Solution : (a) C = C ——> · · · C—C—

(b) − 1 double bond per mer; + 2 single bonds per mer .

Comment : All vinyl polymerizations have the same bond and energy changes; however, the side groups introduce differences in properties.

2-P35 How much energy is released per mer in Problem 2-P34?

Solution : Energy required: + (680,000 J)/(0.6 x10^{24} bonds).
Energy released: − 2(370,000 J)/(0.6 x10^{24} bonds). Net = **−10^{-19} J/mer** .

2-P36 Show how the bonds are altered for the polymerization of propylene to polypropylene.

Answer :
As with all vinyl polymerizations, remove one double bond and create two single bonds.

C = C ——> · · · **C—C—**

Comment : The side **R** does not enter the reaction. However, it does affect the properties.

2-P41 The radius of a K^+ ion is 0.133 nm when CN = 6. (a) What is it when CN = 4? (b) What is it when CN = 8?

Procedure : As footnoted Appendix B (also in Table 2-5.1), the radius increases with more neighbors. $1.1R_4 \cong R_6 \cong 0.97R_8$.

Solutions : (a) $r_{CN=4} = (r_{CN=6})/1.1 = 0.133\text{ nm}/1.1 =$ **0.121 nm** .

(b) $r_{CN=8} = (r_{CN=6})/0.97 = 0.133\text{ nm}/0.97 =$ **0.137 nm** .

2-P42 All the ions in CsI have CN = 8. What is the anticipated center-to-center distance between Cs^+ and the I^- ions?

Procedure : Adjust the radii from App. B for CN = 8. Use ionic radii.

Solutions : $r_{Cs^+} + R_{I^-} = (0.167\text{ nm}/0.97) + (0.220\text{ nm}/0.97) =$ **0.399 nm** .

Also, $d_{Cs^+-Cs^+} = d_{I^--I^-} = a = (0.399\text{ nm})(2)/\sqrt{3} =$ **0.461 nm** .

2-P43 (a) From Appendix B, cite three divalent cations that can have CN = 6 with Se^{2-} ($R_{CN=6}$ = 0.191 nm.), but not CN = 8.

(b) Cite two divalent ions that can have CN = 8 with F^-.

Procedure : For CN = 6: r/R = or > 0.41. For CN = 8: r/R = or > 0.73.

Solutions : (a) r_6 = or > (0.191 nm)(0.41) = 0.078 nm.

r_8 = or < (0.191 nm/0.97)(0.73) = 0.144 nm.

Thus, Ca^{2+}, Mn^{2+}, Hg^{2+}, Pb^{2+}

(b) r_8 = or > (0.133 nm/0.97)(0.73) = or > 0.10 nm.

Thus, Hg^{2+}, Pb^{2+} (possibly Ca^{2+})

2-P44 Why can a neutron move readily through materials, even though the atoms "touch"?

Rationale : The interatomic (and interionic) distances are governed by the local electric fields around the atoms. Neutrons carry no charges, so they can move through these fields unaffected. Positive and negative ions with protons ≠ electrons are not free to penetrate these fields without interference.

2-P45 Strontium is not listed in Appendix B. On the basis of the periodic table (Fig. 2-1.1) and of other data in the appendices, predict the radius of the Sr^{2+} ion.

Prediction : Sr is in Group II. $R_{Be^{2+}} = 0.035$ nm; $R_{Mg^{2+}} = 0.066$ nm;

$R_{Ca^{2+}} = 0.099$ nm; $R_{Sr^{2+}} \approx$ **0.13 nm**

(Actually, $R_{Sr^{2+}} = 0.127$ nm.)

2-P61 Estimate the linear thermal expansion coefficient, α_L, of molybdenum, which has a melting temperature of 2880°C.

Estimation : Based on Fig. 2-6.2, $\alpha_L \approx$ **6 x10^{-6}/°C**. (Actually, $\alpha_L = 5.5 \times 10^{-6}$/°C.)

2-P62 Estimate the elastic modulus of molybdenum based on Fig. 2-6.1.

Estimation : From Problem 2-P61, $T_m = 2880°C$.

Based on Fig. 2-6.1, **$E_{Mo} \approx$ 300 GPa (43,000,000 psi) .**

Actually, E_{Mo} = 330 GPa (47,000,000 psi) .

3-P11 The unit cell of iron is cubic, with a = 0.287 nm. From the density, calculate how many atoms there are per unit cell.

Procedure : Calculate the mass of the unit cell; then divide by the atomic mass.

Solution : $m = \rho V = (7.87 \times 10^6\ g/m^3)(0.287 \times 10^{-9}\ m)^3 = 1.860 \times 10^{-22}\ g$.

$N = (1.860 \times 10^{-22}\ g)/(55.85\ g/0.602 \times 10^{24}\ Fe) =$ **2.00 Fe/u.c.**

3-P12 There are two atoms per unit cell in titanium. What is the volume of the unit cell?

Procedure : Calculate the mass of the unit cell; then divide by the density. Data from App. B .

Solution : $V = [2\ Ti\ (47.90\ amu/(0.602 \times 10^{24}\ amu/g)]/(4.51 \times 10^6\ g/m^3)$

$= 35.3 \times 10^{-30} m^3 =$ **0.0353 nm³**.

3-P13 A piece of aluminum foil is 0.08 mm thick and 670 mm² in area. (a) Its unit calls are cubic with a = 0.4049 nm. How many unit cells are there in the foil? (b) What is the mass of each unit cell? Density is 2.70 Mg/m (= 2.70 g/cm³) ?

Solutions :(a) $(0.08\ mm)(670\ mm^2)/(0.4049 \times 10^{-6} mm)^3 =$ **8.1x10²⁰unit cells**

(b) $\rho = m/V = 2.70 \times 10^6\ g/m^3 = m/(0.4049 \times 10^{-3}\ m)^3$

$m =$ **1.8 x10²² g** (or 107.9 amu) .

3-P21 Calculate the atomic packing factor of vanadium (bcc with a = 0.3039 nm). (Refer to your answers for Problem 247)

Procedure : Divide the volume of two spherical atoms by the volume of the bcc cube. Since the spheres "touch" in a bcc metal, $a = 4R/\sqrt{3}$.

Solution : P.F. $= 2(4\pi/3)R^3/(4R/\sqrt{3})^3 =$ **0.68** .

Comment : We could have used the lattice constant, a , and calculated R = 0.1316 nm. However, with only one type of atom present, the packing factor is independent of the radius.

3-P22 The volume of the unit cell of chromium in Example 3-1.1 is 24 x10⁻³⁰ m³, or 0.024 nm³. Based on the data used in that example problem, plus $PF_{bcc\ metal}$ = 0.68, calculate the radius of the chromium atom to verify the value shown in Appendix B.

Solution : P.F. $= 0.68 = 2(4\pi/3zR^3/0.024\ nm^3$

$R^3 = 0.00195\ nm^3$; $R =$ **0.1249 nm** ; App.B: 0.1249 nm .

[*] These *Practice Problems* either (1) utilize equations directly (plug and chug), (2) refer to specific statements in the text, or (3) "mimic" the examples at the end of the sections. The reader may try them for practice, before attempting the *Test Problems* in the text, which require the student to work from a fuller understanding.

3-P23 From Fig. 3-2.4, show that CN = 12 for an fcc metal.

Reference : Atom at center of the front face. Those in contact are:

4 corner atoms; center-to-center distance $= 2R$.
4 on vertical mid-plane; center-to-center distance $= 2R$.
4 on the vertical mid-plane of the next unit cell in front; c-to-c $= 2R$.

3-P24 (a) How many atoms are there per mm^3 in solid barium?
(b) What is the atomic packing factor?
(c) It is cubic. What is its structure? (Atomic number = 56; atomic mass = 137.3 amu; atomic radius = 0.22 nm; ionic radius = 0.143 nm; density = 3.5 Mg/m^3.)

Procedure : (a) and (b) may be calculated from the density; (c) by observing the P.F.

Calculations : (a)
$$\frac{(3.5 \times 10^{-3}\ \text{g/mm}^3)}{(137.3\ \text{amu/Ba})/(0.602 \times 10^{24}\ \text{amu/g})} = \mathbf{15.3 \times 10^{18}\ Ba/mm^3}$$

(b) $\text{P.F.} = (15.3 \times 10^{18}/\text{mm}^3)(4\pi/3)(0.22 \times 10^{-6}\ \text{mm})^3 = \mathbf{0.68}$.

(c) It is **bcc**, since it is a metal and the packing factor is 0.68 .

3-P25 CsCl has the sc structure of Cl^- ions with Cs^+ ions in the 8-f sites.
(a) The radii are 0.187 nm and 0.172 nm, respectively, for CN = 8; what is the packing factor?
(b) What would this factor be if r/R were 0.73?

Procedure : P.F. = Vol. of ions/Vol. of cube. Solve for a of CsCl from the ionic radii.

(a)
$$\text{P.F.} = \frac{(4\pi/3)[(0.172\ \text{nm})^3 + (0.187\ \text{nm})^3]}{[\,2(0.172\ \text{nm} + 0.187\ \text{nm})/\sqrt{3}\,]^3} = \mathbf{0.684}\ .$$

(b)
$$\text{P.F.} = \frac{(4\pi/3)[(0.73\ \text{nm})^3 + (1\ \text{nm})^3]}{[\,2(0.73\text{nm} + 1\ \text{nm})/\sqrt{3}\,]^3} = \mathbf{0.73}\ .$$

Comment : 0.73 is the maximum packing factor for CsCl-type structures. (Cf. 3-P28.)

3-P26 The intermetallic compound FeTi has the CsCl-type structure with a = 0.308 nm. Calculate its density.

Procedure : The CsCl-type unit cell has 1 Fe and 1 Ti.

Solution : $\rho = [(55.85 + 47.90\ \text{amu})/(0.602 \times 10^{24}\ \text{amu/g})]/[(0.308 \times 10^{-9}\text{m})^3]$
$= \mathbf{5.90\ Mg/m^3}$ (or 5.90 g/cm^3).

3-P27 X-ray data show that the unit-cell dimensions of cubic MgO are 0.412 nm. This material has a density of 3.83 Mg/m^3. How many Mg^{2+} ions and O^{2-} ions are there per unit cell?

Solution :
$$\rho = m/V$$
$$= 3.83\ \text{Mg/m}^3 = \frac{N\,(24.31 + 16.00\ \text{amu})/(0.602 \times 10^{24}\ \text{amu/g})}{(0.412 \times 10^{-3}\text{m})^3}$$

$N_{MgO} = \mathbf{4.00}$, or 4 MgO = (4 Mg^{2+} + 4 O^{2-}).

3-P28 Bunsenite (NiO) has an fcc structure of O^{2-} ions with Ni^{2+} ions in all the 6-f sites. (a) The radii are 0.140 nm and 0.069 nm, respectively; what is the packing factor?
(b) What would this factor be if r/R were 0.41?

Procedure : P.F. = Vol. of ions/Vol. of cube. Solve for a of NiO from the ionic radii.

(a) $$\text{P.F.} = \frac{4\,(4\pi/3)\,[(0.140\text{ nm})^3 + (0.069\text{ nm})^3]}{[\,2(0.140\text{ nm} + 0.069\text{ nm})]^3} = \mathbf{0.71}\ .$$

(b) $$\text{P.F.} = \frac{4\,(4\pi/3)\,[(0.41\text{ nm})^3 + (1.0\text{ nm})^3]}{[\,2(0.41\text{nm} + 1.0\text{ nm})]^3} = \mathbf{0.79}\ .$$

Comment : 0.8 is the maximum packing factor for NaCl-type structures. (Cf. 3-P25.)

3-P29 Lithium fluoride, LiF, has a density of 2.6 Mg/m^3 and an NaCl-type structure. Use these data to calculate the unit cell size, a, and compare it with the value you get from the ionic radii.

Procedure : The NaCl-type cell has 4 cations and 4 anions. AN is Avogadro's No.,0.602 $\times 10^{24}$.

Solutions :
$$V = m/\rho = [4(6.94 + 19.00\text{ amu})/\text{AN}\,]/[2.6\times10^6\text{ g/m}^3]$$
$$= a^3 = 0.0663\times10^{-27}\text{ m}^3; \qquad a = \mathbf{0.405\ nm}\ .$$
$$\text{vs. } a = 2(0.068 + 0.133\text{ nm}) = 0.402\text{ nm}\ .$$

3-P31 From the data for beryllium (hcp) in Appendix B, calculate the volume of the unit cell (Fig. 3-3.2).

Procedure : From Fig. 3-3.2, there are (3 + 2/2 + 12/6), or 6 atoms per unit cell.

Solution :
$$V = m/\rho = [6(9.01\text{ amu})/\text{AN}\,]/[(1.85\times10^6\text{g/m}^3)$$
$$= 0.049\times10^{-27}\text{ m}^3 = \mathbf{0.049\ nm^3}\ .$$

Comments: This is the volume of the hexagonal unit cell. The volume of the rhombic u.c. (Fig. 3-1.1b) is 0.0162 nm^3.

3-P32 What is the ratio of volumes for the two presentations of the hexagonal unit cell in Fig. 3-3.1?

Basis : Triangular prisms.

(a) Hexagonal: 6 prisms. (b) Rhombic: 2 prisms. Therefore, **3/1** .

3-P41 The lattice constant, a, for diamond (Fig. 3-2.6) is 0.357 nm. What percent volume change occurs when it transforms to graphite (ρ = 2.25 Mg/m^3, or 2.25 g/cm^3)?

Procedure : 8 C/unit cell in diamond. Determine the volume of 8 atoms of each phase.

Solution :
$$V_g = m/\rho = 8(12.01\text{ amu AN})/(2.25\times10^6\text{ g/m}^3)$$
$$= 0.0709\times10^{-27}\text{ m}^3 = 0.0709\text{ nm}^3\ .$$
$$V_d = (0.357\text{ nm})^3 = 0.0455\text{ nm}^3;$$
$$\Delta V/V = (0.0709 - 0.0455\text{ nm}^3)/0.0455\text{ nm}^3 = \mathbf{+56\ v/o}\ .$$

Comment : Because of this difference, pressure must be used to make diamonds from graphite.

3-P42 The volume of a unit cell of bcc iron is 0.02464 nm^3 at 912°C. The volume of a unit cell of fcc iron is 0.0486 nm^3 at the same temperature. What is the percent change in density as the iron transforms from bcc to fcc?

Basis :4 atoms = 2 u.c. of bcc, and 1 u.c. of fcc = 4 (55.85 amu) = 223.4 amu .

Solution : ρ_{fcc} = 223.4 amu/0.0486 nm^3 = 4597 amu/nm^3(or 7.64 g/cm^3).

ρ_{bcc} = 223.4 amu/2(0.02464 nm^3) = 4533 amu/nm^3 (or 7.53 g/cm^3).

$(\rho_{fcc} - \rho_{bcc})/\rho_{bcc}$ = (4597 - 4533)/4597 = **+1.4 %** .

Comment : Cf. Example 3-4.1, and Fig. 3-4.1 .

3-P43 MnS has three polymorphs. One of these is the NaCl type structure (Figs. 3-1.1 and 3-2.5). A second is the ZnS-type, which is like diamond (Fig. 3-2.6), except that cations and anions alternate positions (See Problem 329). What percent volume change occurs when the second type (ZnS) changes to the first type (NaCl)?

(See Appendix B for radii where CN = 6, and the footnote explaining what to do when CN ≠ 6.)

Procedure : Each unit cell will have 4 Mn^{2+} and 4 S^{2-} ions. Therefore, we can consider one unit cell of each.

Also, $a_{NaCl} = 2(r + R)$; and $a_{ZnS}\sqrt{3} = 4(r + R)$. Adjust for CN = 4.

Solution : (a) $V_{NaCl} = a^3 = [2(0.080 + 0.184\ nm)]^3 = 0.147\ nm^3$

(b) $V_{ZnS} = a^3 = [4(0.080/1.1 + 0.184\ nm/1.1)/\sqrt{3}]^3 = 0.170\ nm^3$

$(\Delta V/V)_{b\to a} = (0.147 - 0.170\ nm^3)/0.170\ nm^3$ = **-14 v/o** .

Comment : Only the NaCl-type is stable. However, the other two polymorphs can be formed with appropriate starting materials and crystallization procedures.

3-P44 Diamond requires higher pressure to be formed than does graphite. (E.g., it was formed by nature deep in the earth.) What does this suggest about the relative densities of diamond and graphite?

Answer : Pressure produces closer packing, and therefore favors the phase with the greater density. $\rho_d > \rho_g$.

3-P45 From Fig. 3-2.3, show that fcc iron could be categorized as body-centered tetragonal (bct) with a *c/a* ratio of 1.414. Further, the unit cell would be smaller, with only two atoms instead of four. Why do we use fcc rather than bct?

Procedure : Rotate the vertical axis 45°, and set $a_{bct} = a_{fcc}/\sqrt{2}$. Thus, $c = a(1.414)$.

Rationale : Fcc gives greater symmetry. E.g., only one lattice constant is required rather than two. Likewise, there are fewer families of directions and of planes.

3-P51 Refer to Fig. 3-2.1. We can identify the structure of a bcc metal by placing atoms at only two locations. What are they?

Answer : 0,0,0 and $^1/_2,^1/_2,^1/_2$.

(The remaining atoms are redundant translations.)

3-P52 Refer to Fig. 3-2.3. We can identify the structure of an fcc metal by placing atoms at only four locations. What are they?

Answers : $0,0,0$ $\quad$ $^1/_2,^1/_2,0$ $\quad$ $^1/_2,0,^1/_2$ $\quad$ $0,^1/_2,^1/_2$

Comment : Other corners and face centers are integer translations; therefore, redundant.

3-P53 Copper is fcc and has a lattice constant of 0.3615 nm. What is the distance between the 0,1,0 and $^1/_2,0,^1/_2$ locations?

Solution : $0,1,0 \Rightarrow {^1/_2},0,{^1/_2} = \sqrt{(a/2)^2 + (-a)^2 + (a/2)^2}$

$= \sqrt{1.5a^2} = (0.3615\text{ nm})\sqrt{1.5} = \mathbf{0.4427\ nm}$.

3-P54 MnS has the same structure as NaCl (Fig. 3-1.1). Its lattice constant is 0.53 nm. What is the center-to-center distance from the Mn^{2+} ion at $0,0,^1/_2$ and (a) its nearest Mn^{2+} neighbor? (b) its second-nearest Mn^{2+} neighbor?

Solution : (a) $0,0,{^1/_2} \Rightarrow 0,{^1/_2},0 = (a/2)\sqrt{2} = (0.53\text{ nm})/\sqrt{2} = \mathbf{0.375\ nm}$.

(b) $0,0,{^1/_2} \Rightarrow 0,1,{^1/_2} = a = \mathbf{0.53\ nm}$

3-P55 When a copper atom is located at the origin of an fcc unit cell, a small interstitial hole is centered at $^3/_4,^1/_4,^1/_4$. Where are there other equivalent holes within the same unit cell?

Answers : $^3/_4,^3/_4,^3/_4$ $\quad$ $^1/_4,^1/_4,^3/_4$ $\quad$ $^1/_4,^3/_4,^1/_4$

3-P56 Repeat Problem 3-P55. Start with $^1/_4,^3/_4,^3/_4$, rather than $^3/_4,^1/_4,^1/_4$.

Answers : $^1/_4,^1/_4,^1/_4$ $\quad$ $^3/_4,^3/_4,^1/_4$ $\quad$ $^3/_4,^1/_4,^3/_4$

3-P57 Compare the size and "shape" of the holes that are centered (a) at $^1/_2,0,0$ and (b) at $^1/_4,^1/_4,^1/_4$ of an fcc metal.

Procedure : Refer to Fig. 3-2.4(a).

Answers : (a) At $^1/_2,0,0$: CN = **6** ; "diameter" = $\boldsymbol{a - 2R}$.

(b) At $^1/_4,^1/_4,^1/_4$: CN = **4** ; "diameter" = $\mathbf{2[(a/4)\sqrt{3} - R)}$

<u>3-P61</u> (a) A line in the [221] direction passes through the origin. Where does it leave the reference unit cell? (b) Another line in a parallel [221] direction leaves the reference unit cell at 1,1,1. Where did it enter the reference unit cell?

Procedure : Make a sketch.

Answers : (a) $1,1,^1/_2$ $\quad$ (b) $0,0,^1/_2$

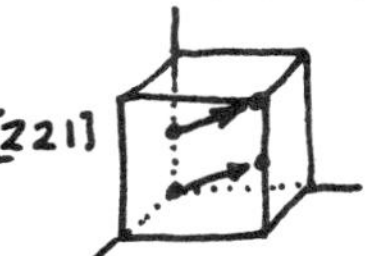

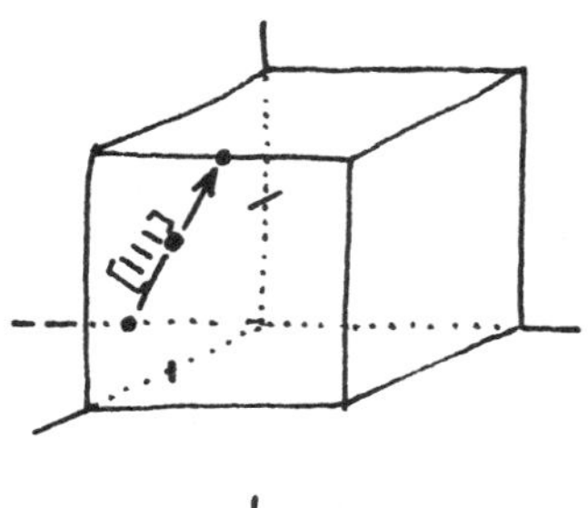

3-P62 A line in the [111] direction passes through location $^1/_2,0,^1/_2$. What are two other locations along its path?

Procedure : Add 1,1,1 or $^1/_2,^1/_2,^1/_2$ etc., to the coefficients of the point $^1/_2,0,^1/_2$.

Solution : $^1/_2,0,^1/_2 + ^1/_2,^1/_2,^1/_2 \Rightarrow 1,^1/_2,1$.

Also, $^1/_2,0,^1/_2 - ^1/_2,^1/_2,^1/_2 \Rightarrow 0,-^1/_2,0$.

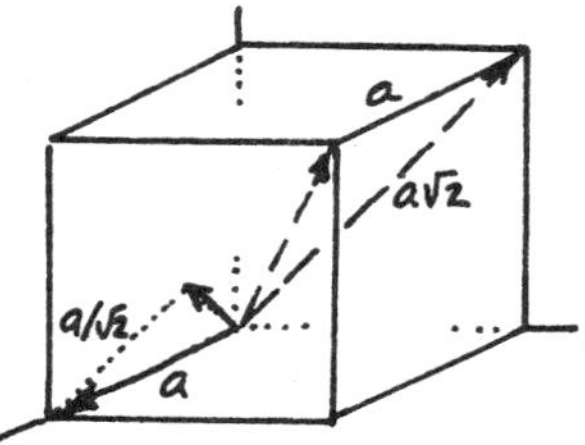

3-P63 (a) In a cubic crystal, what is the tangent of the angle between the [100] direction and the [211] direction? (b) The [011] direction and the [111] direction?

Procedure : Make a sketch.

Solution : (a) $\tan\Theta = (a/\sqrt{2})/a = \mathbf{0.71}$.

(b) $\tan\Phi = a/(a\sqrt{2}) = \mathbf{0.71}$.

3-P64 (a) In a cubic crystal, what is cos [113]∠ [110]? (b) What is sin [010]∠ [1 2 2] ?

Procedure : Make sketches, or use dot product (since it is a cubic crystal).

Solutions : (a) $\cos\Theta = 2/\sqrt{11}\sqrt{2} = 0.43$; $\Theta = \mathbf{64.8°}$

(b) $\cos\Phi = 2/\sqrt{1}\sqrt{9} = 0.67$; $\Phi = 48.2°$. $\sin\Phi = \mathbf{0.75}$.

3-P65 (a) What is the center-to-center spacing of atoms in the <110> directions of copper (fcc, a = 0.361 nm) ?
(b) What is it in the <110> directions of iron (bcc, a = 0.287 nm) ?

Procedure : (a) $a\sqrt{2}$ /(2 spacings)= (0.361 nm)/$\sqrt{2}$ = **0.255 nm** .

(b) $a\sqrt{2}$ /spacing = (0.287 nm)$\sqrt{2}$ = **0.406 nm** .

3-P66 What is the center-to-center spacing of atoms in the [121] direction of copper? (b) What is it in the [121] direction of iron? (See Problem 3-P65 for lattice constants.)

Procedure : Make a sketch from 0,0,0 to 1,2,1 .

Solutions : (a) $a\sqrt{1^2+2^2+1^2}$ /2 spacings

= (0.361 nm/2)$\sqrt{6}$ = **0.442 nm** .

(b) $a\sqrt{1^2+2^2+1^2}$ /1 spacing

= (0.287 nm)$\sqrt{6}$ = **0.703 nm** .

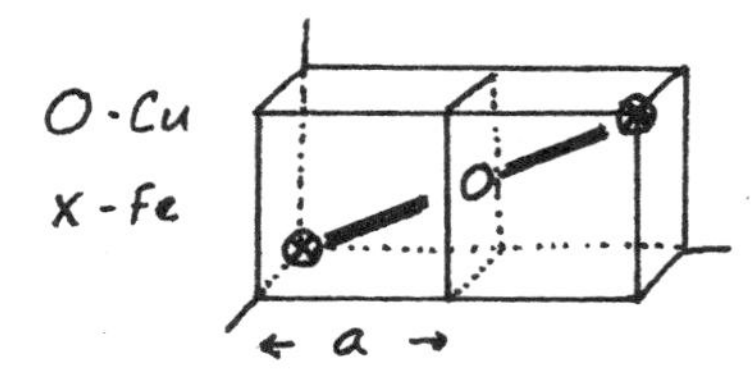

3-P67 What are the several directions in the <012> family of a cubic crystal? (We usually consider [uvw] and [$\overline{u}\overline{v}\overline{w}$] to be of the same direction, but with an opposite sense; that is, [$\overline{u}\overline{v}\overline{w}$] = − [uvw].)

Answers : <012> = { [012], [021], [102], [201], [210], [120]
[01$\overline{2}$], [02$\overline{1}$], [10$\overline{2}$], [20$\overline{1}$], [2$\overline{1}$0], [1$\overline{2}$0] }

Comment : Only a cubic crystal includes all of these directions for the <012> family .

3-P70 A plane intercepts the crystal axes at *a* = 0.5 and b = 0.75. It is parallel to the *z*-axis. What are the Miller indices?
Procedure : Make a sketch .

Answer : Intercepts: **1/2, 3/4,** and ∞
Reciprocals: $(2\ ^{4}/_{3}\ 0)$= (640), or **(320)**

3-P71 A plane intercepts the crystal axes at *a* = 1, *b* = 2 and *c* = 1. What are the Miller indices?
Answer : Intercepts: 1, 2, 1
Reciprocals: $(1\ ^{1}/_{2}\ 1)$, or **(212)**

3-P72 What are the indices for a plane with intercepts at *a* = 1, *b* = $-^{3}/_{2}$, and *c* = $^{2}/_{3}$?
Procedure : Make a sketch .

Answer : Intercepts: 1, -3/2, and 2/3
Reciprocals: $(1\ ^{-2}/_{3}\ ^{3}/_{2})$ = **$(6\bar{4}9)$**

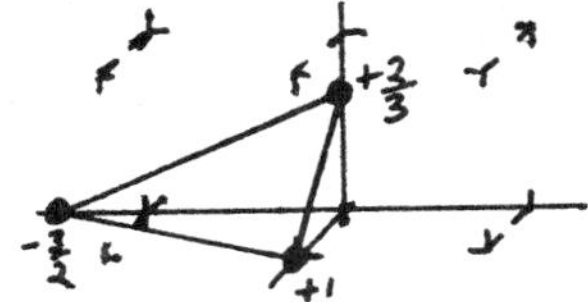

3-P73 (a) How many atoms are there per mm^2 on the (100) plane of copper? (b) How many are there on the (110) plane? (c) How many are there on the (111) plane?
Procedure : Determine the lattice constant; then sketch measurable areas within the planes.
Calculations : $a = 4\ (0.1278\ \text{nm})/\sqrt{2}$ = 0.3615 nm .
(a) (100): $2\ \text{atoms}/a^2 = 2/(0.3615\ \times10^{-6}\text{m})^2$ = **$15.3\ \times10^{12}/\text{mm}^2$** .
(b) (110): $2\ \text{atoms}/a^2\sqrt{2} = \sqrt{2}\ /(0.3615\ \times10^{-6}\text{m})^2$ = **$10.8\ \times10^{12}/\text{mm}^2$** .
(c) (111): $(3/2 + 3/6\ \text{atoms})/[(a\sqrt{2}\ )^2(\sin 60°)/2]$ = $2/(0.866\ a^2)$
= **$17.7\ \times10^{-6}/\text{mm}^2$** .

3-P74 (a) What is the line of intersection between the $(1\bar{1}0)$ and $(\bar{1}12)$ planes of a cubic crystal? (b) What is it between these planes of a tetragonal crystal?
Procedure : Make a sketch, or use the cross product .
Answer : (a) **[110]**
(b) **The same.**

3-P75 (a) What is the line of intersection between the (112) plane and the (100) plane? (b) What is it between the (112) plane and the $(1\bar{1}0)$ plane?
Procedure : Make a sketch, or use the cross product.
Answer : (a) **$[02\bar{1}]$**
(b) $[^{1}/_{2}\,^{1}/_{2}\,^{-1}/_{2}]$, or $[22\bar{2}]$, or **$[11\bar{1}]$**

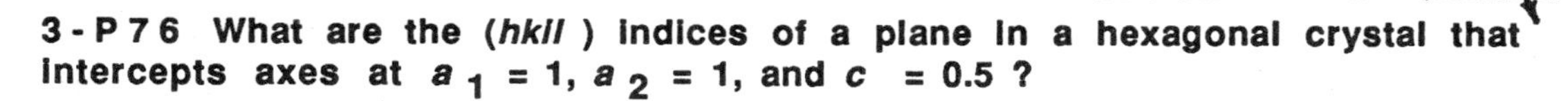

3-P76 What are the (*hkil*) indices of a plane in a hexagonal crystal that intercepts axes at a_1 = 1, a_2 = 1, and *c* = 0.5 ?

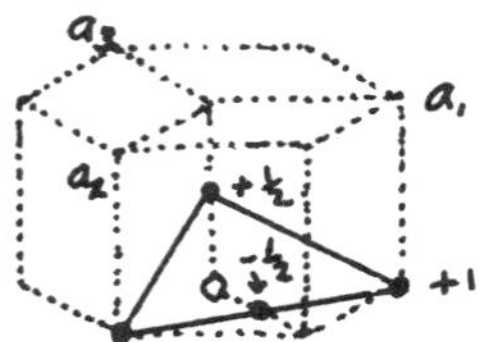

3-P76 (con't) *Procedure* : Make a sketch .
Answer : $a_3 = -0.5$. Therefore, $\mathbf{[11\bar{2}2]}$.
[Check: $h + k = -i$; $1 + 1 = -(-2)$.]

3-P77 List the planes that belong to the {100} family in tetragonal crystals. (b) List those that belong to the {001} family. (c) Identify those that are positive and negative pairs.
Answer : (a) **(100)** **(010)**, and their negatives $(\bar{1}00)$ $(0\bar{1}0)$.
(b) **(001)**, and its negative $(00\bar{1})$.

3-P78 (a) What <110> directions lie in the $(11\bar{1})$ plane of copper? (b) What <110> directions lie in the $(1\bar{1}1)$ plane?
Procedure : Make a sketch .

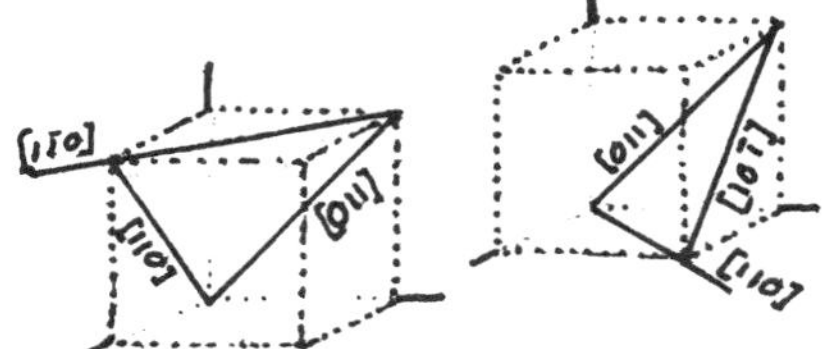

Answers : (a) **[101]**, **[011]**, and $\mathbf{[1\bar{1}0]}$.
(b) **[011]**, **[110]**, and $\mathbf{[10\bar{1}]}$.
Comment : The dot product of indices for the plane and the direction must equal zero. Why?

3-P79 (a) What <111> directions lie in the (101) plane of iron? (b) What <110> directions lie in the $(1\bar{1}0)$ plane?
Procedure : Make a sketch .

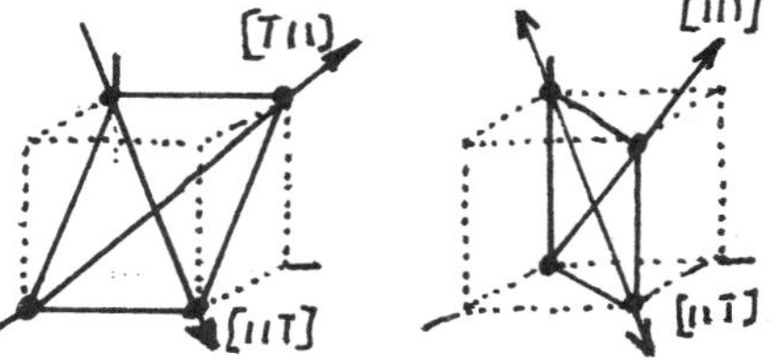

Answers : (a) $\mathbf{[11\bar{1}]}$, and $\mathbf{[\bar{1}11]}$.
(b) **[111]**, and $\mathbf{[11\bar{1}]}$.
Comment : The dot product of indices for the plane and the direction must equal zero. Why?

3-P81 The lattice constant for a unit cell of lead is 0.4950 nm. (a) What is d_{220}? (b) What is d_{111}? (c) What is d_{200}?
Procedure : By inspection from Figs. 3-7.1,3-7.2, and 3-7.3, or by Eq. (3-8.1).
Solutions : (a) $d_{220} = (0.4950 \text{ nm})/\sqrt{2^2 + 2^2 + 0} = \mathbf{0.1750 \text{ nm}}$.
(b) $d_{111} = (0.4950 \text{ nm})/\sqrt{1^2 + 1^2 + 1^2} = \mathbf{0.2858 \text{ nm}}$.
(c) $d_{200} = (0.4950 \text{ nm})/\sqrt{2^2 + 0 + 0} = \mathbf{0.2425 nm}$.

3-P82 A sodium chloride crystal is used to measure the wavelength λ of some x-rays. The diffraction angle 2θ is 27°30' for the d_{111} spacing of the chloride ions. (a) What is the wavelength? (The lattice constant is 0.563 nm.) (b) What would have been the value of 2θ if λ had been 0.058 nm?
Procedure : Use Eq. (3-8.2). Remember that the diffraction angle is 2θ.
Solutions : $d_{111} = (a\sqrt{3})/3 = 0.563 \text{ nm}/\sqrt{3} = 0.325 \text{ nm}$.
(a) $\lambda = 2d \sin\theta = 2(0.325 \text{ nm}) \sin 13.75° = \mathbf{0.154 \text{ nm}}$.
(b) $\sin\theta = \lambda/2d = 0.058 \text{ nm}/2(0.325 \text{ nm}) = 0.089$.
$\theta = 5.1°$; therefore, $2\theta = \mathbf{10.2°}$.

4-P11 The grain boundary area (increases, decreases, remains unchanged) as the grain size increases.

Answer: **Decreases**, because the boundary area, S_V, varies directly with the number of boundary intercepts, and inversely with the mean chord length, $\bar{L}$.

4-P12 Calculate the radius of the largest atom that can fit interstitially into fcc silver without crowding.

Make a sketch: $a = 2r_{hole} + 2R = 4R/\sqrt{2}$

$$r_{hole} = (\sqrt{2}-1)(0.1444 \text{ nm}) = \mathbf{0.06 \text{ nm}}.$$

4-P13 (a) What is the coordination number, CN, of the interstitial site in Problem 4-P12 ?
(b) What structure would result if *every* such site were occupied by a smaller atom or ion?

Answers: (a) CN = **6**. (b) **NaCl** (See Fig. 3-1.1)

4-P14 In copper at 1000°C , one out of every 473 lattice sites is vacant. If these vacancies remain in the copper when it is cooled to 20°C, what will be the density of the copper?

Solution:

$$\rho = \frac{[472]}{[473]} \times \frac{[4(63.54 \text{ g}/0.602 \times 10^{24})]}{[(0.1278 \times 10^{-9}\text{m})(4/\sqrt{2})]^3} = 8.920\times10^6 \text{ g/m}^3, \text{ or } \mathbf{8.920 \text{ g/cm}^3}.$$

(vs. 8.94 g/cm^3, theoretically.)

Conversely: 1/473 of the atoms is missing.

4-P15 Why do ion-pair vacancies form as pairs?

Explanation: The charges must remain balanced. Therefore, **+ = -**.

Comment: This is in AX compounds. Such vacancies would require a 1/2 ratio in AX_2 crystals.

4-P16 The area viewed on a 3 x 4 photomicrograph at ×1000 is what percent of the area viewed in a similar photomicrograph at ×100.

Answer: Magnification is generally expressed in linear dimensions.

Therefore, $(\times100/\times1000)^2 = 0.01$ = **1 percent.**

4-P17 Cite examples of point defects, linear defects, and 2-D defects.

Examples:

Point: Vacancies, interstitials.
Linear: Dislocations, screw and edge.
2-D: Surfaces, grain boundaries.

*** See the footnote with the *Practice Problems* in Chapter 1.**

4-P18 (a) Is the slip vector, b, parellel to (∥) or perpendicular (⊥) to a screw dislocation? (b) What is it to an edge dislocation?

Answer: (See Fig. 4-1.5.) (a) Parallel to a screw dislocation (∥).
(b) Perpendicular to an edge dislocation (⊥).

4-P21 The density of liquid aluminum at its melting point is 2.37 Mg/m^3 Assume a constant atom size; calculate the atomic packing factor.

Procedure : Determine the number of atoms/volume in the liquid; then their "spherical" volume.

Solution : $(2.37 \times 10^6\ g/m^3)/(27\ g/0.602 \times 10^{24}\ Al) = 5.28 \times 10^{28}\ Al/m^3$

P.F. $= [(5.28 \times 10^{28}/m^3)(4\pi/3)(0.14315 \times 10^{-9} m)^3]/[1\ m^3] =$ **0.65**

Alternate : $\rho_L/\rho_{fcc} = [(2.37\ g/cm^3)/(2.70\ g/cm^3)](0.74) =$ **0.65** .

4-P22 Would you expect the heat of fusion of gold to be nearest to 5,000, 10,000, 20,000, 40,000 J/mol)?

Procedure : Refer to the melting temperature. Stronger bonds lead to higher T_m, and ΔH_f .

Estimate : T_m of gold = 1064°C; from Table 4-2.1, ΔH_f is nearest **10,000 J/mole** .

4-P23 Account for the change in slope of the upper curve in Fig. 4-2.4.

Rationale : Below T_g: Expansion from increased vibrational energy only.
Above T_g: Expansion from (1) increased vibrational energy,
and (2) the continuous rearrangements of atoms (or molecules).

4-P24 Why is it easier to form a glass in a polymer than to do so in a metal?

Rationale : Metals crystallize rapidly (ms or µs), because each atom moves individually.
The ordering of molecules into a crystalline array involves the simultaneous rearrangements of large polyatomic groups .

4-P25 Show that the volume expansion coefficient, α_V, approximates three times the linear expansion coefficient, α_L .

Derivation : $V = L^3$, and $(V \pm \Delta V) = (L \pm \Delta L)^3$.

$$1 \pm \Delta V/V = (1 \pm \Delta L/L)^3 = 1 + 3(\Delta L/L) + 3(\Delta L/L)^2 + (\Delta L/L)^3$$

$$\Delta V/V \approx 3(\Delta L/L), \quad \text{or } \alpha_V \approx 3\,\alpha_L$$

4-P30 Is the elastic modulus of cross-linked butadiene rubber greater than, approximately equal to, or less than that of non-vulcanized butadiene rubber.

Answer : Cross-linking anchors adjacent chains to one another. Therefore, the unkinking is less possible as stress is applied. With less strain for a given stress, the elastic modulus is **greater.**

4-P31 Two g of dextrose ($C_6H_{12}O_6$) are dissolved in 14 g of water. What is the number-fraction of each type of molecule?

Sugar solution : Dex.- 2 amu/[2(12) + 12(1) + 6(16) amu/molecule) = 0.011 molecules ;
H_2O - 14 amu/[2(1) + 16 amu/molecule) = 0.778 "

Dextrose = (0.011/0.889) = **0.014**; H_2O = **0.986** .

4-P32 The following data were obtained from an analysis of a polymeric sample:

Interval midpoint	**35,000 amu**	**1.25 g**
" " "	**25,000**	**2.65**
" " "	**15,000**	**2.00**
" " "	**5,000**	**1.90**

Compute the mass-average molecular size.

Solution :

W_i = 1.25 g/7.8 g =	0.160	M_i =	35,000 g/mole	W_iM_i = 5610 g/mole
2.65 /7.8 =	0.340		25,000	8490
2.00 /7.8 =	0.256		15,000	3850
1.90 /7.8 =	0.244		5,000	1220
ΣW_i =	1.000			$\overline{M}_m = \Sigma W_iM_i$ = **19,200** g/mole

4-P33 (a) What is the mass-average molecular size of the molecules in Example 4-3.1 ?
(b) What is the number-average molecular size?

Solution : (a) $\overline{M}_m = \Sigma W_i\, M_i$ (b) $\overline{M}_n = \Sigma X_i\, M_i$

	(a)		(b)	
H_2O:	0.75(18 amu) =	13.5 amu	0.90(18 amu) =	16.2 amu
C_2H_5OH:	0.20(46 ") =	9.2 "	0.094(46 ") =	4.3 "
$C_6H_{12}O_6$:	0.05(180 ") =	9.0 "	0.006(180 ") =	1.1 "
	$M_m = \Sigma W_iM_i$ =	**31.7 amu**	$M_n = \Sigma X_iM_i$ =	**21.6 amu**

Comment : The mass-average molecular weight is always larger than the number-average weight.

4-P34 Polyvinyl chloride $(C_2H_3Cl)_n$ is dissolved in an organic solvent. (a) What is the mean square length of a molecule with a molecular mass of 28,500 g/mol?
(b) What would be the molecular mass of a molecule with one-half the mean square length of that in part (a) ?

Solution : n = (28,500 amu)/[2(12) + 3(1) + 35.5 amu] = 456 mers = 912 bonds.

(a) $\overline{L}$ = (0.154 nm)$\sqrt{912}$ = **4.65 nm** .

(b) $\overline{L}$ = 2.325 nm = 0.154 nm$\sqrt{m}$; m = 228, and n = 114.

Molecular weight = (114 mers/molecule)(62.5 amu/mer)
= **7125 amu/molecule** .

4-P35 Determine the degree of polymerization and the mean-square length of the average molecule in Problem 431, if the polymer is polypropylene. (Number-average molecular weight = 11,900 amu.)

Procedure : We want to consider the *average* molecular weight, and not the individual categories.

Solution : D.P. = (11,900 amu/molecule)/(42 amu/mer) = 283 mer/molecule.

With 2 C—C bonds /mer, m = 566.

$L = l\sqrt{m}$ = (0.154 nm)$\sqrt{566}$ = **3.7 nm** .

4-P36 (a) What percent sulfur would be present if sulfur were used as a cross-link at all possible points in polyisoprene? (b) What percent is used in polychloroprene?

Solution : From Fig. 4-3.10(b).

(a) $(2\ S_2)/(2\ S_2 + 2\ C_5H_8)$ = (2x64)/(2x64 + 2x68) = 0.48, or **48 %.**

(b) $(2\ S_2)/(2\ S_2 + 2\ C_4H_5Cl)$ = (2x64)/(2x64 + 2x88.5) = 0.42, or **42 %.**

Comment : This assumes crosslinks of S_2. Some crosslinks may involve a single sulfur atom.

4-P37 A rubber contains 91 w/o polymerized chloroprene and 9 w/o sulfur. What fraction of the possible cross-links is joined by vulcanization? (Assume that all the sulfur is used for cross-links of the type shown in Fig. 4-3.10.)

Basis : 100 amu rubber = 91 amu chloroprene + 9 amu sulfur.

Sulfur: 9 amu/32 amu = 0.28 sulfurs Chl: 91amu/88.5 amu = 1.03 chl.

Required: 2 S/chl. mer; 0.28 S/2.06 S(req'd) = **0.14** .

4-P38 Sketch three of the four possible isomers of butanol (C_4H_9OH) .

Sketches :

```
  H  H  H  H            H  H  H
H C- C- C- C-OH       HC- C- C H
  H  H  H  H           H  |  H          H  OH H
                          HCH          HC- C- CH
  H  H  H  H               |            H  |  H
H C- C- C- C H            OH              HCH
  H  H  OH H                               H
```

4-P39 Sketch the structure of the various possible isomers for octane, C_8H_{18} .

Sketches (only the carbons are shown):

```
C-C-C-C-C-C-C-C    C-C-C-C-C-C     C-C-C-C-C-C     C-C-C-C-C-C       C C         C C
                       C C           C     C           C           C-C-C-C-C    C-C-C-C
C-C-C-C-C-C-C                              C           C                 C        C C
            C      C-C-C-C-C-C     C-C-C-C-C-C
                       C   C              C
C-C-C-C-C-C-C                                      C-C-C-C-C          C
        C          C-C-C-C-C-C           C           C C C        C-C-C-C-C
                       C C         C-C-C-C-C-C                         C
C-C-C-C-C-C-C                              C           C C             C
      C                                            C-C-C-C-C
                                                       C
```

4-P41 Name the components of four common substitutional alloys.

Brass: Zinc in copper. Stainless steel: Chromium (and nickel) in iron.

Bronze: Tin in copper. 12 carat gold: 50 % gold (balance, commonly copper).

Sterling silver: Copper in silver. Monel: Copper in nickel (App. C).

4-P42 An alloy contains 85 w/o copper and 15 w/o tin. Calculate the a/o of each element.

Basis : 100 amu alloy = 85 amu Cu + 15 amu Sn .

Cu: 85 amu/(63.54 amu/Cu atom) = 1.338 Cu **91.4 a/o Cu;**

Sn: 15 amu/(118.7 amu/Sn atom) = 0.126 Sn **8.6 a/o Sn.**

Comment : For heavier solutes, a/o are less than w/o.

4-P43 There is 5 a/o magnesium in an Al-Mg alloy. Calculate the w/o magnesium.

Basis : 100 atoms alloy = 5 atoms Mg + 95 atoms Al .

Al: 95 atoms(26.98 amu/Al atom) = 2563.10 amu Al **95.5 w/o Al;**

Mg: 5 atoms(24.31 amu/Mg atom) = 121.55 amu Mg **4.5 w/o Mg .**

Comment : For lighter solutes, a/o are greater than w/o.

4-P44 Consider Fig. 4-4.3 to be an interstitial solution of carbon in fcc iron. What is the w/o carbon present?

Basis : 32 atoms Fe + 2 atoms C (as sketched).

Fe: 32 atoms (55.85 amu/ Fe atom) = 1787 amu Fe **98.7 w/o Fe;**

C: 2 atoms (12.01 amu/ C atom) = 24 amu C **1.3 w/o C .**

Comment : Most steels contain less than 0.8 w/o carbon.

4-P45 Consider Fig. 4-4.2 to be a substitutional solid solution of copper and gold. (a) What is the w/o Cu present if Cu is the more prevalent atom? (b) What is the w/o Cu present if Au is the more prevalent atom?

Basis : (a)10 atoms Au + 22 atoms Cu (as sketched).

Au: 10 atoms(197.0 amu/Au atom) = 1970 amu Au **58.5 w/o Au;**

Cu: 22 atoms(63.54 amu/Cu atom) = 1397.9 amu Cu **41.5 w/o Cu .**

(b) The same calculation, but with Au >Cu: **87.2 w/o Au;** **12.8 w/o Cu .**

4-P46 Carbon atoms can be dissolved into the largest interstices of fcc iron. (a) How many of these sites (per unit cell) are available? (b) How many neighboring iron atoms surround these sites?

Answer: (a) **4 sites/u.c.** (Cf. Na^+ in NaCl, Fig. 3-1.1.)

(b) **6-f sites.**

4-P47 Based on the data in Appendix B, which metallic element among Al, Au, Cd, Ni, and Zn should have the greater solubility in solid copper?

Factors : Size, similarity of crystal structure, electrical similarity (valence).

Nickel R_{Ni} = 0.1246 nm vs. 0.1278 for Cu; both prefer fcc; both can form 2^+ cations.

4-P48 Based on the data in Appendix B, which will have the smallest unit cell dimension,-- brass, bronze, or pure copper?

Answer : **Pure Cu.** Zn and Sn are both larger than Cu; therefore, they will expand the lattice as they enter into solid solution with copper.

Comment : In a *random* solid solution, the radius may be averaged on the basis of atom percent.

4-P51 (a) What is the w/o FeO in the solid solution of Fig. 4-5.1? (b) What is the w/o Fe^{2+}? (c) What is the w/o O^{2-}?

Solution : Based on Fig. 4-5.1: 10 Fe^{2+}, 17 Mg^{2+}, 27 O^{2-}, 10 FeO, and 17 MgO .

(a) (10 FeO)(55.85 + 16.00 amu) = 718.5 amu FeO **51 w/o FeO ;**

(17 MgO)(24.31 + 16.00 amu) = 685.3 amu MgO **49 w/o MgO .**

(b) (10 Fe^{2+})(55.85 amu)/(718.5 + 685.3 amu) = 0.398 **39.8 w/o Fe^{2+} .**

(c) (27 O^{2-})(16.00 amu)/(718.5 + 685.3 amu) = 0.308 **30.8 w/o O^{2-}.**

4-P52 If all the iron ions of Fig. 4-5.1 were changed to Co^{2+} ions, what would be the w/o MgO?

Solution : (10 CoO)(58.93 + 16.00 amu) = 749.3 amu CoO **52 w/o CoO ;**
(17 MgO)(24.31 + 16.00 amu) = 685.3 amu FeO **48 w/o MgO .**

Comment : We normally label compound solid solutions in terms of "molecules," even though the atoms do not pair off as molecules.

4-P53 A solid solution contains 30 m/o MgO and 70 m/o LiF. (a) What are the w/o of Li^+, Mg^{2+}, F^-, and O^{2-}? (b) What is the density?

Basis : 100 cations + 100 anions (= 25 unit cells).

Mg^{2+}:	30(24.31 amu) =	729 amu	= **24 w/o**	r = 0.3(0.066) + 0.7(0.068)
Li^+:	70(6.94 amu) =	483 amu	= **16 w/o**	= 0.0674 nm
O^{2-}:	30(16.00 amu) =	480 amu	= **16 w/o**	R = 0.3(0.140) + 0.7(0.133)
F^-:	70(19.00 amu) =	1330 amu	= **44 w/o**	= 0.1351 nm
		3022 amu		

ρ = [(3022 g/0.602x10^{24})/(25u.c.)]/[2(0.0674 +0.1351)(10^{-7}cm)]^3 = **3.02 g/cm^3.**

4-P61 A copolymer has a 5-to-2 mer ratio of styrene and butadiene. What is the weight ratio of these to components?

Solution :
Styrene- $5(C_2H_3C_6H_5)$ 5 x 104 amu = 520 amu
Butadiene- $2(C_4H_6)$ 2 x 54 amu = 108 amu
= **4.8 to 1 .**

Comments : This copolymer contains both types of mers in the same chain (Fig. 4-6.1(b). This is in contrast to the mutual solubility shown in Fig. 4-6.1(a)between PS and PPO.

4-P62 Polyvinyl chloride and polyvinylidene chloride (Table 2-3.1) are copolymerized in a 2-to-1 weight ratio. They form a molecular chain similar to that of Fig. 4-6.1. What fraction of the mers is of each type?

Basis: 30,000 amu = 20,000 amu PVC + 10,000 amu PVDC

PVC: 20,000 amu/(24 + 3 + 35.5 amu) = 320.0 f_{PVC} = **0.76**

PVDC: 10,000 amu/(24 + 2 + 71 amu) = 103.1 f_{PVDC}= **0.24**

5-P11 A syrup contains equal quantities of water and sugar. How much more sugar can be dissolved into 100 g of the syrup at 80°C?
Procedure : Make a materials balance. Sugar in the parts = sugar in the whole. Sol.Lim. = 79%.
Solution : (0.50)(100 g) + x = (0.79)(100 + x); x = **138 g** added sugar†.

5-P12 A molten lead-tin solder has an eutectic composition. Assume 50 g are heated to 200°C. How many g of tin can be dissolved into this solder?
Solution : 100 g solder + x g added tin; solubility limit = 0.75 tin.
Tin balance: (50 g)(0.619) + x =0.75(50 + x g); x = **26.2 g Sn.**

5-P13 One ton of salt (NaCl) is spread on the streets after a winter storm. The temperature is -15°C (5°F). How many tons of ice will be melted by the salt?
Data : From Fig. 5-1.2, the solubility limit = 0.82 H_2O, and therefore 0.18 NaCl.
Solution : NaCl balance: 0.18(1 T + x) = 1 T; x = **4.6 T ice melted**•.
Comment: The resulting brine can contain no more than 82 w/o ice.

5-P21 A 90Cu-10Sn bronze (Fig. 5-6.2) is cooled slowly from 1100°C to 20°C. What phase(s) will be present as the cooling progresses?
Procedure : Follow the 90-10 bronze composition down the temperature scale.
Answer: L from 1100 -> 1010°C; (L + α) from 1010 -> 830°C;
α from 830 -> 340°C; (α + ϵ)† from 340 -> 20°C.

5-P22 A 65Cu-35Zn brass (Fig. 5-6.1) is heated from 300°C to 1000°C. What phase(s) are present at each 100°C interval?
Answer: From Fig. 5-6.1, we can simply read the phase fields that are crossed.
300, 400, 500, 600, 700°C; α. 800*, 900°C; (α + β). 1000, 1100°C; **Liquid.**

5-P23 An alloy contains 90Pb-10Sn. (a) What phases are present at 100°C, 200°C, and 300°C? (b) Over what temperature range(s) will there be only one phase?
Procedure : From Fig. 5-2.1, locate the one-phase areas for a 90-10 alloy.
Answer: (a) 100°C, (α + β); 200°C, α; 300°C, α (admittedly a slight amount) + **L.**
(b) Only α, **150 to 270°C**; only liquid, **above 305°C.**

5-P24 Locate the solidus on the Cu-Zn diagram (Fig. 5-6.1).
Answer: (Fig. 5-6.1) It lies immediately below all of the (x + L) fields, with horizontal lines at 903, 835, 700, 598, and 424°C.

• Color displacement during printing may produce a slight variance (± 0.5 %)

5-P25 Show the sequential changes in phases when the composition of an alloy is changed from 100% Cu to 100% Al (a) at 700°C, (b) at 450°C, (c) at 900°C. (See Fig. 5-6.3.)

Answer[•]: 700°C (%Al):

Phase	Range	Phase	Range
α	(0>8%)	$\gamma + \epsilon_2$	(21>23)
$\alpha + \beta$	(8>11%)	ϵ_2	(23>24)
β	(11>12.5)	$\epsilon_2 + L$	(24>36)
$\beta + \gamma$	(12.5>15)	L	(36>100%)
γ	(15>21)		

450°C:

Phase	Range	Phase	Range
α	(0>9%);	$\zeta + \eta_2$	(26>28)
$\alpha + \gamma_2$	(9>16);	η_2	(28>29)
γ_2	(16>20),	$\eta_2 + \theta$,	(29>46.5)
$\gamma_2 + \delta$	(20>21),	θ	(46.5>47)
δ	(21-21.5)	$\theta + \kappa$	(47>97.5)
$\delta + \zeta$	(21.5>24.5)	κ	(97.5>100)
ζ	(24.5>26)		

900°C:

Phase	Range	Phase	Range
α	(0>7.5)	$\gamma_1 + \epsilon_1$	(18>21)
$\alpha + \beta$	(7.5>10)	ϵ_1	(21>22)
β	(10>14)	ϵ_1+Liq.	(22>25.5)
$\beta + \gamma_1$	(14>15.5);	Liq.	(25.5>100)
γ_1	(15.5>18)		

5-P26 The solidus temperature is an important temperature limit for hot-working processes of metals. Why?

Explanation : The solidus is the upper limit for hot-working a metal, if the presence of a liquid is to be avoided. Liquid has the potential for introducing "hot-shortness" into the metal,--meaning rupturing during deformation.

5-P27 (a) Show the sequential changes in phases when the composition of an alloy is changed from 100% Al to 100% Mg at 300°C. (b) Show the changes at 400°C. (See Fig. 5-6.4.)

Answer[•]: 300°C: (%Mg):

Phase	Range	Phase	Range
α	(0>6%)	$\beta' + \gamma$	(42>49)
$\alpha + \beta$	(6>35)	γ	(49>57)
β	(35>37)	$\gamma + \epsilon$	(57>94)
$\beta + \beta'$	(37>40$^+$	ϵ	(94>100%) .
β'	(40$^+$>42)		

400°C:

Phase	Range	Phase	Range
α,	(0>11%)	γ	(46>59)
$\alpha + \beta$,	(11>35)	$\gamma + \epsilon$	(59>91)
β,	(35>37)	ϵ	(91>100%).
$\beta + \gamma$	(37>46)		

[•] Color displacement during printing may produce a slight variance (± 0.5 %)

5-P31 **Refer to Fig. 5-6.4. (a) At 500°C, what is the solubility limit of magnesium in solid ∝ ? (b) What is it in liquid ? (c) What are the chemical compositions of the phase(s) in a 40Al-60Mg alloy at 500°C ? (d) What are they in a 20Al-80Mg alloy ?**

Procedure : If in a 2-phase field, read the answer directly from the solubility curve at the end of the tie line. In a 1-phase field, the phase composition is identical with the alloy composition.

Answers †: (a) **11% Mg** (and 89% Al). (b)**76% Mg** (and 24% Al).

(c) Liquid: **40Al-60Mg** (not saturated). (d) Liq.: **24Al-76Mg**; ϵ: **7.5Al-92.5Mg.**

Comment : It is convention to report the compositions of condensed phases in weight percent (w/o), unless stated otherwise. Atom percents, a/o, are shown along the top abscissa.

5-P32 **Refer to Fig. 5-2.3. (a) At 2000°C, what is the solubility limit of Al_2O_3 in the liquid? (b) What is it in β ? (c) What are the chemical compositions of the phase(s) in a 20Al_2O_3-80ZrO_2 ceramic at 1800°C? What are they at 2000°C?**

Answers : (a) **78%** Al_2O_3. (c) ∝: **98** Al_2O_3-**2** ZrO_2; β: **4** Al_2O_3-**96**ZrO_2.

(b) **6%** Al_2O_3. (d) Liquid: **42** Al_2O_3-**58** ZrO_2 ; β: **6**Al_2O_3-**94** ZrO_2.

5-P33 **What are the chemical compositions of each phase in Problem 5-P22?**

Answers †: 300, 400, 500, 600, 700°C: ∝ (65Cu-35Zn);

800°C: ∝(66Cu-34Zn)*,and β (61Cu-39Zn);

900°C: ∝(67.5Cu-32.5Zn),and β (63Cu-37Zn);

1000 °C: **Liquid** (65 Cu-35 Zn); 1100 °C: **Liquid** (65 Cu-35 Zn).

5-P34 **What are the chemical compositions of each phase in Problem 5-P21?**

Answers †: (Fig. 5-6.2)

Temp.°C	∝: Cu-Sn	ϵ: Cu-Sn	Liq.: Cu-Sn
1100	-	-	90 -10
1000	98 - 2	-	89 -11
900	95 - 5	-	81 -19
800-400	90 - 10	-	-
300	94 - 6	62.8-37.2	-
200	98 - 2	62.8-37.2	-
100	Near 100-0	62.8-37.2	-

5-P41 **A 90Al-10Mg alloy (Fig. 5-6.4) is melted and then is cooled slowly. (a) At what temperature does the first solid appear? (b) At what temperature does it have 2/3 liquid and 1/3 ∝? (c)At what temperature does it have 1/2 liquid and 1/2 ∝? (d) At what temperature does it have 99+% ∝ with a trace of liquid?**

Procedure : (a) calls for the liquidus; (d), the solidus. Interpolate for (b) and (c).

Answer: (a) **620°C.** (b) **600°C.** (c) **590°C.** (d) **520°C.**

---

† Color displacement during printing may produce a slight variance (± 0.5 %)

5-P42 **Repeat Problem 5-P41, but for a 10Al-90Mg alloy (ϵ for α).**

Answer: (a) **600°C.** (b) **575°C.** (c) **560°C.** (d) **465°C.**

5-P43 **What composition of Ag and Cu will possess (a) $^1/_4$ α and $^3/_4$ β at 600°C? (b) What composition will possess $^1/_4$ liquid and $^3/_4$ β at 800°C?**

Procedure: The composition is at the center of gravity for the amounts of the two phases.

Answer: (a) **26Ag-74Cu.** (b) **23Ag-77Cu.**

5-P44 **What composition of Al_2O_3 and ZrO_2 will possess (a) $^3/_4$ α and $^1/_4$ β at 1800°C? (b) What composition will possess $^3/_4$ liquid and $^1/_4$ β at 1950°C?**

Procedure: The composition is at the center of gravity for the amounts of the two phases.

Answer: (a) **$75Al_2O_3$-$25ZrO_2$.** (b) **$40Al_2O_3$-$60ZrO_2$.**

5-P45 **(a) At what temperature will a monel alloy (70% nickel, 30% copper) contain $^2/_3$ liquid and $^1/_3$ solid? (b) What will be the composition of the liquid and of the solid?**

Procedure: (a) Interpolation on the **1370°C (2500°F)** isotherm.

(b) Liquid: **67Ni-33Cu.** α: **76Ni-24Cu** .

5-P46 **Assuming 1500 g of 90-10 bronze in Problem 5-P21, what are the masses of solid phase(s) at each 100°C interval?**

Solution:

1100°C:	No solid		
1000 :	α: 1/(11-2)(1500 g)	= **170 g.**	
900 :	α: (19-10)/(19-5.5)(1500 g)	= **1000 g.**	
800 - 400 :	Only α.	= **1500 g.**	
300 :	α: (~0.9)(1500 g)	= **1350 g**	ϵ = 150 g
200 :	α: (~0.76)(1500 g)	= **1150 g**	ϵ = 350 g
100 :	α: (~1.0)(1500 g)	= **1500 g**	

5-P47 **At 175°C, how many g of α are there in 7.1 g of eutectic Pb-Sn solder?**

Solution: α = (99 - 61.9)/(99 - 17) = 0.45; 0.45(7.1 g) = **3.2 g.**

5-P48 **With 200 g of 65Cu-35Zn, how many grams of α are present at each temperature of Problem 5-P22?**

Answer: 300 -> 700°C, **200 g**; 800°C, **~180 g**; 900°C, **~100 g**; 1000°C, **none.**

5-P49 **At what temperature can an alloy of 40Ag-60Cu contain 55% β?**

Procedure: Use a transparent mm-scale to interpolate on Fig. 5-2.5. The bottom of the (β+L) field has too little β (50% β); the top of the (α+β) field has too much β (62% β). Thus, 55% β is realized when the temperature crosses the eutectic at **780°C.** Three phases coexist at this invariant temperature (Eq 5-2.2).

5-P51 Write the equation for the eutectoid reaction that is in the $BaTiO_3$-$CaTiO_3$ system (Fig. 5-6.7).

Equation: β (24% $CaTiO_3$) $\xrightarrow[105°C]{}$ α (20% $CaTiO_3$) + γ(91% $CaTiO_3$).

5-P52 Locate four eutectoids in the Cu-Sn system (Fig. 5-6.2). (b) Write the reactions for two of these.

Equations:

During cooling —

β (25% Sn) $\xrightarrow[586°C]{}$ α (15.8% Sn) + γ(25.5% Sn).

γ (27% Sn) $\xrightarrow[520°C]{}$ α (15.8% Sn) + δ(32% Sn).

δ (32.6% Sn) $\xrightarrow[350°C]{}$ α (11% Sn) + ϵ(37% Sn).

ζ (34% Sn) $\xrightarrow[580°C]{}$ δ(33% Sn) + ϵ(37% Sn).

5-P53 The phenol-aniline system possesses two eutectics. Write an equation for each, using the format of Eq. (5-2.1) with w/o.

Procedure: Same as previously; but note that the lower abscissa is m/o, with w/o above.

Equations:

L (21 w/o An) $\xrightarrow[15°C]{}$ α (0 w/o An) + β(49.7 w/o An).

L (90 w/o An) $\xrightarrow[-12°C]{}$ β(49.7 w/o An) + γ(100 w/o An).

5-P54 A "monotectoid" is not listed among the several invariant reactions of Section 5-5. Why?

Explanation: If L_3 of Eq. (5-5.6) were replaced with a solid, S_3, to give an "-oid" reaction, the reaction would match the **eutectic** reaction: $L_2 \xrightarrow[\text{cooling}]{} S_1 + S_3$.

Comment: Alternatively, the replacement of L_2 in Eq. (5-5.6) with S_2 would be possible; however, it is sufficiently rare so that it has not been labeled. It produces a liquid during cooling!!

(There is one unusual situation of this type in the Cu-Sn system. Can you locate it?)

5-P61 An alloy of 50 g Cu and 30 g Zn is melted and cooled slowly. (a) At what temperature will there be 40 g α and 40 g β ? (b) At what temperature will there be 50 g α and 30 g β ? (c) At what temperature will there be 30 g α and 50 g β ?

Solution: This is a 62.5-37.5 alloy. By interpolation on Fig. 5-6.1:

(a) **780°C** . (b) **750^{+}°C** . (c) **800^{+}°C** .

5-P62 How much mullite will be present in a 60% SiO_2-40% Al_2O_3 brick (10 lb) at the following temperatures under equilibrium conditions? (a) 1400°C. (b) 1580°C. (c) 1600°C.

From Fig. 5-2.4 :

(a) 1400°C (10lb)(40-0)/(71-0) = **5.6 lb mullite**

(b) 1580°C (10lb)(40-0)/(71-0) = **5.6 lb mullite**

(c) 1600°C (10lb)(40-5.5)/(71-5.5) = **5.3 lb mullite**

6-P11 The polymerization of ethylene releases energy (Example 2-3.2). This indicates that C_2H_4 is less stable than is $\text{-}(C_2H_4)\text{-}_n$. Why is it that ethylene can remain as a less stable phase almost indefinitely?
Explanation : It takes energy (+) to break the C=C double bonds, before the C—C— single bonds release energy (-) .

6-P12 Name a common candy product that is a glassy solid.
Examples : Rock candy, etc.
Comments : Refer to Fig. 5-1.1 . A boiling syrup can contain up to 80% sugar. If this is cooled rapidly, as with peanut brittle, it does not have time for the sugar to crystallize. The syrup becomes an amorphous solid. In contrast, a nucleant (cream of tartar) is commonly added to fudge so that many fine sugar crystals form, rather than fewer coarse crystals. This gives a "creamy" fudge, in contrast to a "sugary" candy.

6-P13 Sterling silver possesses 7.5 w/o copper. Why is it that amount?
Reference : Fig. 5-2.5. 7.5 w/o Cu was at the practical limit for processing a single-phase alloy when the composition was officially defined. At 8.5%, for example, there would be only a 15°C range with α alone. Any overshoot would produce some liquid, as could a slight variation in the Cu content. (Today's process control technology could permit slightly more than 7.5% Cu.)

6-P14 Metallic glasses must be quenched from their liquid to ambient temperatures in milliseconds or less. The cooling rates for silica and polymeric glasses can be $\ll$1°C/s without crystallization. Why is there a difference?
Explanation : In metals, the necessary rearrangement into a long-range crystal structure is atom-by-atom, each more or less independently. In silica glasses (Fig. 2-4.3), considerable energy is required to break the covalent bonds and thereby introduce rearrangements. Also, large molecules require more time to be maneuvered into the long-range crystalline structure.

6-P21 Consider a 40MgO-60FeO composition. There is a progressive increase in the FeO content of the liquid as it solidifies. The FeO content of the solid also increases as the temperature drops (Fig. 5-6.6). How can both phases increase in FeO without an overall compositional change?
Explanation : Refer to Fig. 5-6.6. During solidification, the *amount* of liquid (x >60% FeO) decreases, and the *amount* of solid (x < 60% FeO) increases. The total product [(% FeO x amount)$_L$ + (%FeO x amount)$_S$] remains constant.

6-P22 Assume 20 kg of an 8Al-92Mg alloy are first melted and then cooled rapidly to 500°C. Since there is no time for diffusion in the solid, the average composition, $\bar{\epsilon}$, of the solid is 5% Al. (a) What is the composition of the liquid at 500°C ? (Assume rapid diffusion in the liquid.) (b) How much liquid is there at 500°C ? (c) At what temperature will the second solid phase appear? (d) What is the composition of the final liquid?

6-P22 (con't) *Solution* : (a) **24Al-76Mg** .

(b) Since $\overline{\epsilon}$ has 5 % Al, [(8 - 5)/(25 - 5)](20 kg) = **3 kg Liquid** .

(c) **437°C** (d) **32 Al-68 Mg** .

6-P23 A long rod of silicon is "zone-refined" by being moved slowly through a short heater such that the molten zone moves from one end and toward the other end. It solidifies as it leaves the heated zone. Explain how this process could be repeated to remove impurities.

Explanation : Consider Si (+Al) as an example (Fig. 5-3.3).

The solidifying silicon (β) is almost pure. The aluminum in this example remains in the liquid. which is carried to the end of the rod. Discard the Al-rich end.

Still greater purification is possible, because a repeat pass will concentrate the Al once again into the last liquid to solidify.

6-P31 Refer to Example 6-3.1. At 975°C, the free energy value for the reaction, ΔF_V, is -0.8×10^9 J/m³. On a volume basis, how much larger must the nuclei be than they are at 900°C?

Solution : From Eq. (6-3.3), $r_c = -2\gamma/\Delta F_V$

$$r_c = -2(0.5 \text{ J/m}^2)/(-0.8 \times 10^9 \text{ J/m}^3) = 1.25 \times 10^{-9} \text{ m.}$$

Or **1.25 nm** (about "10 atomic radii") involving approximately 1000 atoms.
Thus, compared to Example 6-3.1, a **15-to-1** volume ratio.

6-P32 Nucleation may develop at various microstructural sites more readily than by the homogeneous route involved in Example 6-3.1. What are these sites:

Sites :
i) Foreign particles, e.g. AgI for cloud "seeding;"
ii) external surfaces;
iii) grain boundaries;
iv) dislocations;
v) solute atoms.

6-P41 An aluminum wire is stretched between two rigid supports at 35°C. It cools to 15°C. What additional stress is developed?

Data : From App. C. $\alpha_L = 22.5 \times 10^{-6}$/°C.

Solution : $\Delta L/L = 0 = \alpha_L(-20°C) + s/E$.

$$s = -(-20°C)(70{,}000 \text{ MPa})(22.5 \times 10^{-6}/°C) = \textbf{31 MPa.}$$

or

$$s = -(-20°C)(10^7 \text{ psi})(22.5 \times 10^{-6}/°C) = \textbf{4,500 psi} .$$

6-P42 (a) Estimate the linear expansion coefficient of bcc iron at 900°C from the data in Fig. 3-4.1
(b) Why does this value not match the data of Appendix C?

Procedure : Draw a tangent to the α-Fe curve at 900°C.

Solution : (a) $\alpha_L \cong (0.014)/(900 - 150°C)$, or **~18 x10⁻⁶/°C** .

(b) The data in App. C are for **20°C.** Normally, α_V increases with temperature because of the asymmetry of the energy trough (Fig. 2-5.3).

6-P43 Which will be higher, the *mean energy* or the *median energy* of the gas molecules in your room?

Answer: **Mean > median energy.** This is the general case for skewed distributions (Figs. 6-4.1 and 6-4.2) because the tail of the range has a longer moment arm.

6-P44 Refer to Example 6-4.2. What fraction of the atoms have sufficient energy to jump out of their sites at 1000°C?

Solution : $\ln(n/N) = -2.92 - (0.214 \times 10^{-18}\ \text{J})/(13.8 \times 10^{-24}\ \text{J/K})(1273\ \text{K})$;

$n/N = 2.8 \times 10^{-7} = \mathbf{1/(3.6 \times 10^{6}\ atoms)}$.

6-P45 An activation energy of 2.0 eV (or 0.32 $\times 10^{-18}$J) is required to form a vacancy in a metal. At 800°C, there is one vacancy for every 10^4 atoms. At what temperature will there be one vacancy for every 1000 atoms?

Procedure : Energy requirements follow the Arrhenius ($\ln x$ vs $1/T$) relationship.

Solution: $(n/N)_{800} = 10^{-4} = M \exp[-0.32 \times 10^{-18}\ \text{J}/(13.8 \times 10^{-24}\text{J/K})(1073\ \text{K})]$

$(n/N)_T = 10^{-3} = M \exp[-0.32 \times 10^{-18}\ \text{J}/(13.8 \times 10^{-24}\text{J/K})\,T\,]$

Solving simultaneously, T = 1201 K, or **928°C** ; (and $M = 2.4 \times 10^5$) .

6-P51 A solid solution of copper in aluminum has 10^{26}atoms of copper per m^3 at point X, and 10^{24}copper atoms per m^3 at point Y. Points X and Y are 10 μm apart. What will be the diffusion flux of copper atoms from X to Y at 500°C ?

Procedure : Diffusivity = (4×10^{-14} m^2/s) at 500°C (Table 6-5.1)

Solution : $J = -(4 \times 10^{-14}\ \text{m}^2/\text{s})[10^{24} - 10^{26}/\text{m}^3)/10^{-5}\ \text{m}]$

$= \mathbf{4 \times 10^{17}/m^2{\cdot}s}$.

6-P52 (a) What is the ratio of diffusivities for carbon in bcc iron to carbon in fcc iron at 500°C?

(b) What is it for carbon in fcc iron to nickel in fcc iron at 1000°C?

(c) Of carbon in fcc iron at 1000°C to carbon in fcc iron at 500°C?

(d) Why are the ratios greater than one?

Data: Table 6-5.1. These can be calculated for other temperatures from Table 6-5.2

Answers: (a) $D_{\text{C in bcc Fe}}/D_{\text{C in fcc Fe}} = (10^{-12}\ \text{m}^2/\text{s})/(5 \times 10^{-15}\ \text{m}^2/\text{s}) = \mathbf{200/1}$.

(b) $D_{\text{C in fcc Fe}}/D_{\text{Ni in fcc Fe}} = (3 \times 10^{-11}\ \text{m}^2/\text{s})/(2 \times 10^{-16}\ \text{m}^2/\text{s}) = \mathbf{150{,}000/1}$.

(c) $D_{1000°\text{C}}/D_{500°\text{C}} = (3 \times 10^{-11}\ \text{m}^2/\text{s})/(5 \times 10^{-15}\ \text{m}^2/\text{s}) = \mathbf{6{,}000/1}$.

(d_a) Bcc has a lower **PF** than fcc. (d_b) $r_C < R_{Ni}$. (d_c) More **thermal energy** at 1000°C.

6-P53 The inward flux of carbon atoms in fcc iron is 10^{19}/m^2·s at 1000°C What is the concentration gradient?

Procedure : This is the reverse of Problem 6-P51; but calculating for the gradient.

Solution : $J = 10^{19}/\text{m}^2{\cdot}\text{s} = -(3 \times 10^{-11}\ \text{m}^2/\text{s})(\Delta C/\Delta x)$. $(\Delta C/\Delta x) = \mathbf{-3.3 \times 10^{29}/m^4}$

Comment: The gradient is negative in the direction of flow.

6-P54 **(a) Using the data of Table 6-5.2, calculate the diffusivity of copper in aluminum at 400°C. (b) Check your answer against Fig. 6-5.4. Do the values match?**

Solution : (a) $D = (0.15 \times 10^{-4}\ m^2/s)\exp[(-0.210 \times 10^{-18}\ J)/(13.8 \times 10^{-24}\ J/k)(673\ K)]$

$= \mathbf{2 \times 10^{-15}\ m^2/s}$

(b) $\log_{10}(2 \times 10^{-15}) = (-15+0.3)$. **Checks OK.**

6-P55 **A zinc gradient in copper alloy is 10 times greater than the aluminum gradient in a copper alloy. Compare the flux of solute atoms/m²·s in the two alloys at 500°C. (The data for $D_{Al\ in\ Cu}$ are given in Example 6-5.3.)**

Solution : $J_{Zn\ in\ Cu} = -(4 \times 10^{-18}\ m^2/s)(10\ G)$

$J_{Al\ in\ Cu} = -(2.5 \times 10^{-17}\ m^2/s)(G)$ $J_{Zn\ in\ Cu}/J_{Al\ in\ Cu} = \mathbf{1.6}$.

6-P56 **Aluminum is to be diffused into a silicon single crystal. At what temperature will the diffusion coefficient be 10^{-14} m²/s? (Q = 73,000 cal/mol and D_o = 1.55 x 10^{-4} m²/s.)**

Solution : $\ln 10^{-14} = \ln(1.55 \times 10^{-4}\ m^2/s) - (73{,}000\ cal/mol)/(1.987\ cal/mole{\cdot}K)T$.

T = 1566 K, or **1293°C** .

Comment : Much of the chemical literature still expresses energy as cal/mol.

6-P57 **Refer to Table 6-5.1. (a) Why are the values higher for couple 2 than for couple 1? (b) Why are they higher for couple 2 than for couple 4? (c) Why are they higher for couple 11 than for couple 10? (d) Why are they higher for couple 8 than for couple 9?**

Answers : (a) $PF_{bcc} < PF_{fcc}$. (c) $PF_{boundary} < PF_{xtal}$.

(b) $r_C < R_{Fe}$. (d) Al-Al bond is weaker than Cu-Cu bond. (Cf. T_m.)

6-P58 **Refer to Problem 6-P51. (a) What is the diffusion coefficient of copper in aluminum at 100°C? (b) What will be the diffusion flux of copper atoms from X to Y at 100°C?**

Solution : (a) $D = (0.15 \times 10^{-4}\ m^2/s)\ \exp[-0.210 \times 10^{-18} J/(13.8 \times 10^{-24} J/K)(373\ K)]$

$= \mathbf{3 \times 10^{-23}\ m^2/s}$.

(b) $J = -(3 \times 10^{-23}\ m^2/s)[10^{24}-10^{26}/m^3)/10^{-5}\ m] = 3 \times 10^{8}/m^2{\cdot}s = \mathbf{300/mm^2{\cdot}s}$.

6-P59 **(a) In Al_2O_3, what is the $D_{Al^{3+}}/D_{O^{2-}}$ ratio at 2000 K? (b) Give two reasons for this significant difference.**

Procedure : Take the data from Fig. 6-5.5, being careful to interpolate on a $\log_{10}$ basis.

Answer : (a) $D_{Al^{3+}}/D_{O^{2-}} = (10^{-14.8}\ m^2/s)/(10^{-17.5}\ m^2/s) = \mathbf{500}$.

(b) The O^{2-} ions are much **larger** than the Al^{3+} ions. Also, because of the 3/2 valance ratio, 1/3 of the cation sites are **vacant** and available for diffusion steps.

7-P11 Estimate the grain-boundary area per unit volume for the iron in Fig. 4-1.8. The magnification is ×500.

Sample	Points per unit length	P_L =	S_V =
Perimeter:	15/[4(55 mm/500)]	34/mm	~70 mm²/mm³
Diagonals:	12/[2(77.5 mm/500)]	39/mm	80
38-mm circle:	10/[π(38 mm/500)]	42/mm	80
Center "axes":	11/[2(55 mm/500)]	50/mm	100
Average of above	48/(600 mm/500)	40/mm	**~80 mm²/mm³**

Comment: Each of these is a sample. As demonstrated, second samplings will provide somewhat different, but useful values.

7-P12 (a) Assume that the ASTM G.S.#6 of Fig. 7-1.6 represents a two-dimensional cut through a polycrystalline solid. Estimate the corresponding grain-boundary area.
(b) Repeat part (a) for G.S. #2.

Estimate : ×100. Use a 25-mm (1.0 in.) circle. (a) Three samplings gave P_L= 24, 23, and 21.

$$P_L = (\sim 23)/(25\pi \text{ mm}/100) \approx 30/\text{mm}; \qquad S_V \approx \mathbf{60\ mm^2/mm^3}.$$

(b) $$S_V = 2P_L = 2(\sim 6)/(25\pi \text{ mm}/100) \approx \mathbf{15 mm^2/mm^3}, \quad \text{or } \mathbf{380\ in.^2/in.^3}.$$

7-P13 With ASTM G.S.#6, how many times as much grain-boundary area is there as with ASTM G.S. #3.

Solution : $N_6/N_3 = 2^5/2^2 = 32/4 = 8 = [1/(\overline{L}_6)^2]/[1/(\overline{L}_3)^2]$

$[1/\overline{L}_6]/[1/\overline{L}_3] = \sqrt{8} = P_6/P_3 = (2S_6)/(2S_3);$ S_6/S_3 **= 2.8** .

7-P14 Grain size may increase noticeably at higher temperatures. Describe what happens with respect to grain size at lower temperatures.

Answer : The grains continue to grow but at a much *reduced* rate.

7-P21 Cite the geometric variables that exist (a) in single-phase microstructures, and in (b) multiphase microstructures.

Variables : (a) 1) Grain size, 2) grain shape, and 3) grain orientation.
(b) (Grain size, grain shape, grain orientation), plus 4) phase fraction, and 5) phase distribution.

7-P22 An alloy of 95Al-5Cu is solution-treated at 550°C, then cooled rapidly to 400°C where it is held for 24 hours. During that time it produces a microstructure with 10^6 particles of θ per mm^3. The θ particles ($CuAl_2$) are nearly spherical and have approximately 60% greater density than the matrix.

(a) Approximately how far apart are these particles?
(b) What is the average particle dimension?

7-P22 (con't) *Procedure* : Assume cubic distribution. Calculate the average distance, $\overline{D}$, bewteen the particles. From the volume fraction and the total volume (θ + κ), calculate the volume and the average diameter, $\overline{d}$, of the particles.

Basis : From Fig. 5-6.3, (98-95)/(98-46) = 5.8 w/o θ . 100 g alloy = 5.8 g θ + 94.2 g

(a) 1 particle/10^{-6}mm^3 = 1particle/(10 μm)^3 = **10 μm** between particles.

(b) [5.8 g /(1.6 P_κ g/cm^3)]/[5.8 g /(1.6P_κ) + 94.2/P_κ] = 0.037, or 4 v/o.

With 10^{-6}mm^3 (total volume) per particle,

$$V_{particles} = 0.037(10^{-6} mm^3) = (\pi/6)\delta^3$$

$$\delta = 0.004 \text{ mm} = \mathbf{4\ \mu m.}$$

7-P23 A microstructure has nearly spherical particles of β with an average dimension $\overline{d}$ that is 10% of the average distance $\overline{D}$ between the centers of the adjacent particles. (a) What is the volume percent of β? (b) What is the ratio $\overline{d}/\overline{D}$ with 0.5 v/o β ?

Procedure : Assume 1 particle per cube, $\overline{D}^3$, each with a volume of $(\pi/6)\overline{d}^3$.

(a)Vol. fract.$_{\overline{d}/\overline{D}}$ = 0.1: $(\pi\overline{d}^3/6)/\overline{D}^3$ = 0.52 $(\overline{d}/\overline{D})^3$ = 0.52(0.1)^3= 0.0005 = **0.05 v/o.**

(b) $\overline{d}/\overline{D}_{0.5 \text{ v/o}}$: 0.52 $(\overline{d}/\overline{D})^3$ = 0.005; $\overline{d}/\overline{D}$ = **0.2**

Comment : Had we assumed a bcc pattern of particles, rather than cubic, Vol. frac. = 0.68 $(\overline{d}/\overline{D})^3$, rather than 0.52 $(\overline{d}/\overline{D})^3$. With an fcc pattern, Vol. frac. = 0.74 $(\overline{d}/\overline{D})^3$.

7-P24 An Al-Cu alloy contains 2.5 v/o θ (ρ = 4.4 Mg/m^3) in a matrix of κ which is essentially pure Al. What is the density of the alloy?

Procedure : Based on Eq. (7-2.2), $\rho_m = \Sigma f_i \rho_i$

Solution : ρ_m = 0.025(4.4 Mg/m^3) +0.975(2.7 Mg/m^3) =2.74 Mg/m^3, or **2.74 g/cm^3.**

7-P25 Why do curves for isothermal reactions involving supercooled phases commonly have a "C" shape, such as in Fig. 7-2.2?

Rationale : Slight supercooling: Nucleation rate $\dot{N}$ is slow; therefore, $\dot{N}\dot{G}$ is low; and t is long.

Major supercooling: Diffusion is slow; growth rate, $\dot{N}\dot{G}$ is slow and t is long.

Intermediate supercooling: $\dot{N}$ and $\dot{G}$ are moderate; $\dot{N}\dot{G}$ is faster, and t is shorter.

Write the eutectoid reaction found in the $BaTiO_3$-$CaTiO_3$ system (Fig. 5-6.7).

Equation : β (24 $CaTiO_3$) $\xrightarrow{105°C}$ α(20 $CaTiO_3$) + γ(91 $CaTiO_3$) .

7-P32 (a) Locate three eutectoids in the Cu-Al system that contain less than 20 % Al (Fig. 5-6.3). (b) Write the reactions for two of these eutectoids.

Equations : β(11.5 Al) $\xrightarrow{565°C}$ α(9 Al) + γ_2(15 Al) .

(Reactions for cooling)

γ_1(15 Al) $\xrightarrow{780°C}$ β (13 Al) + γ_2(15.5 Al) .

χ (15.5 Al) $\xrightarrow{965°C}$ β(14.5 Al) + γ_1(16 Al) .

7-P33 (a) What phases are present in a 99.8 Fe-0.2 C steel at 800°C ? (b) Give the compositions of these phases. (c) What is the fraction of each?
Answers: Refer to Fig. 7-3.1.
(a) α and γ (b) α: **Fe (+ <0.01% C)**; γ: **99.7 Fe-0.3 C**
(c) By interpolation: 0.2/0.3 = **66%** γ, and **34%** α.

7-P34 Refer to Fig. 5-4.2. Make a similar presentation of phase fractions for 10 g of a 79 $BaTiO_3$-21$CaTiO_3$ ceramic (Fig. 5-6.7).

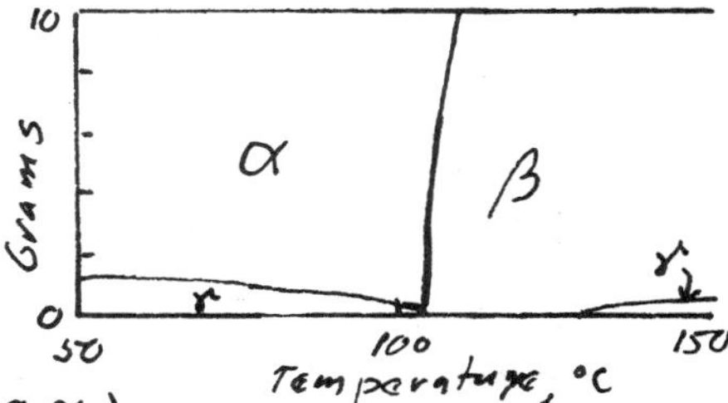

Interpolation : 50°C: (10/85)(10 g) = 1.2 g γ, (and 8.8 g α)
Repeat for 104°C, 106°C, and 150°C.

7-P35 The maximum solubility of carbon in ferrite (α) is 0.02 w/o. Show by calculation how many unit cells there are per carbon atom?
Basis: 10,000 amu = 9998 amu Fe + 2 amu C.
C: 2 amu /(12.01 amu/C) = 0.1665 C \
Fe: 9998 amu/(55.85 amu/Fe) = 179 Fe = 89.5 u.c. / 1 C per **540 unit cells** .

7-P36 Iron carbide has a density of 7.6 Mg/m^3(= 7.6 g/cm^3). Its unit cell is orthorhombic (Section 3-1), and contains 12 iron atoms plus 4 carbon atoms. What is its volume ?
Solution: V = [12(55.85) + 4(12.01 amu)]/[(0.602 x10^{24} amu/g)(7.6 x10^6 g/m^3)
= 160 x10^{-30} m^3 = **0.160 nm^3** = abc, where $a \neq b \neq c$.

7-P37 Describe the phase changes that occur on heating a 0.45% carbon steel from room temperature to 1200°C. (Ferrite, α; austenite, γ; carbide, $\overline{C}$.)
Reference: Fig. 7-3.1.
R.T. $\xrightarrow{\alpha + \overline{C}}$ 727°C; (α + $\overline{C}$ + γ) at 727°C; 727°C $\xrightarrow{\alpha + \gamma}$ 770°C; 770°C $\xrightarrow{\gamma}$ 1200°C.
Comment: When 3 phases are present, the temperature is fixed,-- 727°C. There is a range of temperatures for 2-phase mixtures. See the Phase Rule (Section 5-3).

7-P38 Without referring to an Fe-Fe_3C phase diagram, indicate the compositions of the phases in Problem 7-P37 at selected temperatures. (Knowing these, you will find Test Problems easier to solve.)
For checking : Fig.7-3.1.
α: 0% at 20°C to 0.02% at 727°C to 0% at 912°C.
$\overline{C}$: Fe_3C has 6.7% C at all temperatures of interest.
γ: Approx. 0.8% at 727°C; dropping to 0.45 % at 770°C; 0.45% above 770°C.
Also, label the α + $\overline{C}$; the α + γ, and the γ + $\overline{C}$ fields.

7-P39 Why do eutectoid reactions generally require longer times than do polymorphic reactions. For example, compare Eq. (5-5.4) and (3-4.2)?

Explanation : In polymorphic reactions, the reactant and product are identical in composition. A eutectoid reaction produces two new phases with compositions that are distinct from the initial phase. As a result diffusion is required over distances up to a mm or more. The atoms move only nanometer distances in polymorphic reactions as the atoms are rearranged into new structures.

8-P11 A 7.6-mm (0.30-in.) diameter 1040 steel bar, which was initially 2.27 m (89.4 in.) long, supports a weight of 3963 N (890 lb_f). What is the difference in length if the steel bar is replaced by a 70-30 brass bar?

Procedure : Determine the applied stress. Then, use the elastic modulus to calculate the strains.

Solution : s = 3963 N/[(π/4)(0.0076 m)^2] = 87.4 MPa .

e_{steel} = 87.4 MPa/205,000 MPa = 0.00042;

e_{brass} = 87.4 MPa/110,000 MPa = 0.00079;

ΔL = (0.00079 − 0.00042)(2.27 x 10^3 mm) = **0.84 mm.**

or s = 890 lb_f/[(π/4)(0.30 in.)^2] = 12,600 psi .

e_{steel} = 12,600 psi/30,000,000 psi = 0.00042;

e_{brass} = 12,600 psi/16,000,000 psi = 0.00079;

ΔL = (0.00079 − 0.00042)(89.4 in.) = **0.033 in.**

8-P12 What is the volume change of a rod of brass when it is loaded axially by a force of 233 MPa (33,800 psi)? (Poisson's ratio is 0.37.)

Procedure : Determine the axial strain; then the two lateral strains for Poisson's ratio.

Solution : e_z = 233 MPa/110,000 MPa = 0.0021;

$e_x = e_y$ = -(0.37)(0.0021) = −0.0008 (contraction);

(1 + 0.0021)(1 − 0.0008)^2 = 1.0005 $\Delta V/V$ = **0.05 v/o.**

Alternatively, $\Delta V/V$ = 0.0021 +(−0.0008) + (−0.0008) = 0.0005, or **0.05 v/o.**

Comment : English units give the same strains: e_x = 33,000 psi/16x10^6 psi = 0.0021, etc.

8-P13 What is the bulk modulus of the brass in Problem 8-P12? The shear modulus?

Procedure : *K, G, and E* are related through Poisson's ratio Eqs. (8-1.4) and (8-1.5).

Solution :(a) $K = E/3(1-2\upsilon)$ = 110 GPa/3(1−0.74) = **140 GPa**

or = 16,000,000 psi/3(0.26) = **20,500,000 psi** .

(b) $G = E/2(1 + \upsilon)$ = 110 GPa/2(1.37) = **40 GPa**, or 16 Mpsi/2(1.37) = **5.8 Mpsi.**

8-P14 If iron has an axial stress of 208 MPa (30,200 psi), what will be the highest local stress within a polycrystalline iron bar?

Procedure : With only elastic starin, the strain in all grains will be identical. Since $s = e\,E$, the highest local stress will be in the directions within grains with the highest elastic modulus.

Solution : $\bar{s}/s_{max} = \bar{E}\,e/E_{max}\,e$; s_{max} = 208 MPa (280 GPa/205 GPa) = **284 MPa** ;

or s_{max} = 30,200 psi (41 Mpsi/30 Mpsi) = **41,000 psi** .

8-P15 A precisely machined steel rod has a specified diameter of 18.6 mm (0.732 in.). It is to be elastically loaded longitudinally with a force or 670,000 N (150,000 lb_f). (a) By what percentage will its diameter change? (b) By what percentage will the in cross-sectional area change? (Poisson's ratio is 0.29.)

Solution : A = (π/4)(0.0186 m)^2 = 272 x10^{-6} m^2, or 272 mm^2 .

(a) $e_z = s/E = (F/A)/E$ = [(670,000 N)/(272 x10^{-6} m^2)]/205 x10^9 Pa = 0.012 .

e_z = -(0.29)(0.012) = -0.0035; $\Delta d/d$ = **-0.35 %.** d_e = 18.54 mm.

(b) $\Delta A/A$ = (π/4)[(0.01854 m)^2-(0.01860 m)^2]/[(π/4)(0.0186 m)^2)] = **-0.7%**

Alternatively : 1 + ΔA = (1 - 0.0035)^2; ΔA = -0.007, or **-0.7%** .

8-P16 The elastic modulus of copper drops from 110 GPa (16×10^6 psi) at 20°C (68°F) to 107 GPa (15.5×10^6 psi) at 50°C (122°F). What is the change in the total length of a 1575-mm (62-in.) bar if the stress is held constant at 165 MPa while the temperature rises those 30°C.

Procedure : Assume that the load is removed at 20°C, and then reapplied at 50°C.

Solution : $-\Delta L/L$ at 20°C (to be recovered) = −165 MPa/110,000 MPa = −0.00150.

(Zero load at 20°C to zero load at 50°C) + elastic strain at 50°C.

$\Delta L/L$ = $(17 \times 10^{-6}/°C)(+30°C)$ + (165 MPa/107,000 MPa) = 0.00205

Therefore, ΔL = (0.00205 - 0.00150 mm/mm)(1575 mm) = **0.87 mm** ;

or ΔL = (0.00205 - 0.00150 in./in.)(62 in.) = **0.034 in.**

8-P17 Based on Problems 8-P12 and 8-P13, what shear angle, α, does a shear stress of 262 MPa (38,000 psi) produce in brass?

Procedure : Shear strain, γ, is the tangent of the displacement angle, α.

Solution : γ = 262 MPa/40,000 MPa = 0.0066 = tan α; α = **0.38°** ;

or γ = 38,000 psi/5,800,000 psi = 0.0066 = tan α; α = **0.38°**

8-P18 Distinguish among the three elastic moduli.

Young's modulus:	Axial stress/axial strain	$E = s/e$.
Shear modulus:	Shear stress/ shear strain	$G = \tau/\gamma = \tau/\tan\alpha$.
Bulk modulus:	Hydrostatic pressure/volume change	$K = P_h/(\Delta V/V)$.
	or $[\text{compressibility}]^{-1}$	$K = 1/\beta$

8-P21 "The yield strength is the stress where strain switches from elastic to plastic." Comment on the validity of this statement.

Comment: Although plastic strain, e_{pl}, starts with a critical stress called the yield strength ,S_y, elastic strain starts with the first applied stress, because even the initial forces stretch the interatomic bonds. Bond stretching continues beyond S_y as higher and higher stresses are applied. Above S_y, strain is the sum of the two; $e = e_{el} + e_{pl}$. In many situations, $e_{el} \ll e_{pl}$; so the elastic deformation is overlooked, but still present and continues to increase with larger loads.

8-P22 A wire of a magnesium alloy is 1.05 mm (0.04 in.) in diameter. Plastic deformation starts with a load of 10.5 kg (which is _____N), or 23 lb_f. The total strain is 0.0081 after loading to 12.1 kg (26.6 lb_f).

(a) How much permanent strain has occurred with a load of 12.1 kg (26.6 lb_f)?

(b) Rework this problem with English units.

Procedure : Since plastic deformation starts with a load of 10.5 kg (102.9 N), there must be both elastic and plastic deformation at 12.1 kg; thus, $e = e_{el} + e_{pl}$ = 0.0081. We can calculate e_{el} from the elastic modulus and the force of $(12.1 \text{ kg})(9.8 \text{ m/s}^2)$ = 118.6 N.

Solution : e_{el} = $[(118.6 \text{ N})/(\pi/4)(0.00105 \text{ m})^2]/(45 \times 10^9 \text{ N/m}^2)$ = 0.0030

e_{pl} = 0.0081 - 0.0030 = **0.005** .

Or e_{pl} = $0.0081 - [(26.6 \text{ lb}_f)/(\pi/4)(0.04 \text{ in.})^2]/(6.5 \times 10^6 \text{ psi})$ = **0.005** .

8-P23 A brass test specimen has a reduction of area of 35 percent. (a) What is the true strain, ϵ? (b) What is the ratio of the true stress, σ, to the nominal stress, s, for fracture, S_b?

Procedure : The fracture area, A_f , is 65 percent of the original area, A_o . See Eq. (8-2.4).

Solution : (a) $\epsilon = ln\,(1.0\,A_o/0.65\,A_o) =$ **0.43** .

(b) $\sigma/s = (F/A_{true})/(F/A_{nominal}) = 1.0/0.65 =$ **1.54** .

8-P31 Identify the 12 $<\bar{1}11>\{101\}$ slip systems for a bcc metal.

Answer :

$[11\bar{1}](101)$	$[1\bar{1}1](110)$	$[11\bar{1}](011)$
$[\bar{1}11](101)$	$[\bar{1}11](110)$	$[1\bar{1}1](011)$
$[1\bar{1}1](10\bar{1})$	$[11\bar{1}](\bar{1}10)$	$[111](01\bar{1})$
$[111](10\bar{1})$	$[111](\bar{1}10)$	$[\bar{1}11](01\bar{1})$

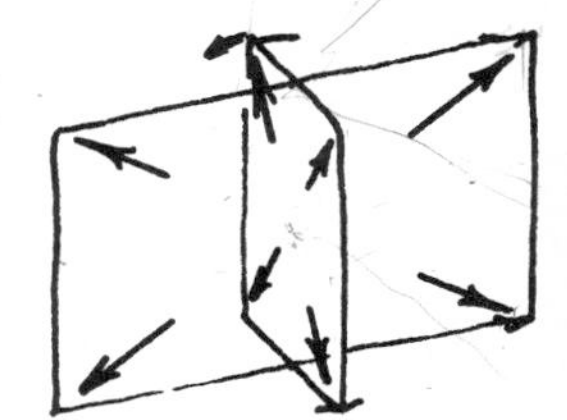

8-P32 (a) Sketch the atomic arrangement on the (110) plane of MnS, which has the structure of NaCl (Fig. 3-2.5).
(b) Identify the six $<1\bar{1}0>\{110\}$ slip systems for this compound.
(c) How long is the slip vector, b, in these systems?

Sketch : (a)

```
S  —— S  —— S
|     |     |
Mn —  Mn —  Mn
|     |     |
S  —— S  —— S
```

(b)

$[1\bar{1}0](110)$
$[110](\bar{1}10)$
$[10\bar{1}](101)$
$[101](10\bar{1})$
$[01\bar{1}](011)$
$[011](0\bar{1}1)$

(c) $a = 2(0.080 + 0.184\text{ nm}) = 0.528\text{ nm}$;
$b = 0.528\text{ nm}(\sqrt{2}/2) =$ **0.373 nm** .

8-P33 Differentiate between the terms, *resolved shear stress*, and *critical shear stress*.

Resolved shear stress: **Axial stress, *s* , resolved into the shear component, τ.**

Critical shear stress : **The shear stress on the slip plane that initiates slip.**

Comment : We can derive $\tau = s\ \cos\lambda\ \cos\Phi$, where λ is the angle between the force direction and the shear direction, and Φ is the angle between the force direction and the slip plane normal.

8-P34 The engineer commonly defines yield strength as the critical stress to produce a 0.2% offset on the *s-e* curve (Fig. 8-2.1b). Based on Fig. 8-1.5, why is this offset necessary?

Rationale : -Slip is not identical in each grain. i) Some grains are oriented for higher resolved shear stresses. (The highest is at 45° to the force direction, and zero at 0° and 90°.)
ii) Some grains are oriented for higher elastic moduli (Fig.8-1.5); thus, higher stresses develop.

-The net effect is plastic strain starts gradually, and progressively. Initial detection depends on the analytical equipment. An 0.2% offset is a defined amount that is tolerable in most engineering design. If required, smaller offsets may be specified.

8-P35 Alloys of zinc and of magnesium are more widely used as casting alloys than as wrought (plastically deformed) alloys. Suggest a valid reason why this is so.

Rationale : Both metals are hexagonal. As indicated in Table 8-3.1, very few slip systems are available, and deformation by twinning is limited.

8-P36 Refer to Table 8-3.2. Why are there <100>{001} slip systems in CsCl structures, and not <111>{$\bar{1}\bar{1}0$} slip systems? There are more atoms per mm in the <111> directions than in the <100> directions.

Make a sketch :

Slip requires displacement to an identical lattice site. $b_{111} = a\sqrt{3}$; $b_{100} = a$.

The slip vector is longer in the <111> direction.

The required energy is proportional to b^2.

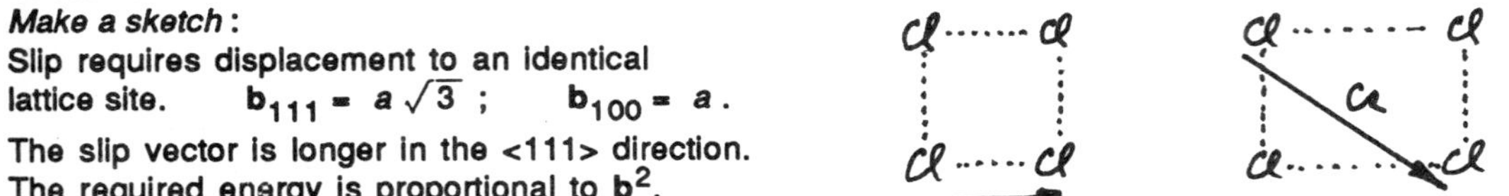

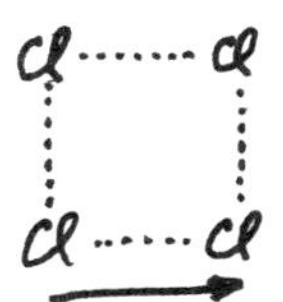

8-P41 What would the follow through angle have been in Example 8-4.1 if, before testing the sample, the initial angle of the pendulum had been set at 105°?

Procedure : The fracture energy (toughness) equals the change in the pendulum's potential energy.

Solution : $\Delta E = -36.8\text{ J} = [\cos(-105°) - \cos\theta][10\text{ kg}\ (9.8\text{m/s}^2)(0.75\text{ m})]$

$\cos\theta = 0.24; \qquad \theta = 76°.$

8-P42 The value of K_{Ic} for a steel is 186 MPa·m$^{1/2}$ (169,000 psi·in.$^{1/2}$). What is the maximum tolerable crack when the steel carries a nominal stress of 800 MPa (116,000) psi? (The geometric factor, Y, is 1.1 .)

Reference : Equation (8-4.1) .

Solution : $186\text{ MPa}\cdot\sqrt{\text{m}} = 800\text{ MPa}\ (1.1)\sqrt{\pi c}$. $\qquad c = 0.014\text{ m} = 14\text{ mm};$

or $\qquad 169{,}000\text{ psi}\cdot\sqrt{\text{in.}} = 116{,}000\text{ psi}(1.1)\sqrt{\pi c}$. $\qquad c = 0.56\text{ in.}$

8-P43 Convert K_{Ic} = 135,000 psi·in.$^{1/2}$ to MPa·m$^{1/2}$.

Conversion : $(135{,}000\text{ psi}\cdot\sqrt{\text{in.}})(6.894\times10^{-3}\text{ MPa/psi})/\sqrt{39.37\text{ in/m}}$

$= 148.3\ \text{MPa}\cdot\sqrt{\text{m}}$. Thus, $1\text{ MPa}\cdot\sqrt{\text{m}} = 910\text{ psi}\cdot\sqrt{\text{in.}}$

8-P44 The testing machine of Fig. 8-4.2 is described in Example 8-4.1. What is its maximum capacity expressed in the amount of energy absorbed by a specimen?

Procedure : Assume the pendulum is stopped by the test specimen, and all the energy is absorbed.

Solution : $E = (10\text{ kg})(9.8\text{ m/s}^2)(0.75\text{ m})[\cos(-120°) - \cos 0°] = -110\text{ J}.$

$E_{max} = -110$ J (energy lost by the pendulum) = **+110 J** (absorbed by the specimen).

8-P45 A small hole drilled through a steel plate ahead of a crack will stop the crack's progress until repairs can be made. Explain how the hole accomplishes this feat.

Rationale : The stress concentration is lower ahead of the drill hole (which has no crack) than at the tip of the original crack. Alternatively, the blunt crack (the drill hole) does not concentrate the stress as much as a sharp crack.

Comment : The drill hole is not a repair. However, it does buy time for repairs. Stresses are still concentrated ahead of the hole. As a result there is a greater probability that a new crack will start there than elsewhere along the edge.

9-P11 Based on the data of Section 4-2, what is the maximum percent porosity that could exist in a magnesium casting, if no riser were used?

Procedure : This will be the percent volume shrinkage from liquid to solid.

Solution : From Fig. 4-2.2, $V_{liq.} = 1.00 + 0.11$; $V_{sol.} = 1.00 + 0.06$.

$\Delta V/V = (1.06 - 1.11)/1.11 = -0.045$, or **4.5 v/o porosity** from shrinkage.

9-P12 A 1.00-in. x 0.25-in. aluminum bar (commonly sold in a hardware store) was extruded from a 100-lb, 6-in. diameter aluminum billet. Assuming 1 percent end scrap, how many feet of product were obtained? (1 g/cm^3 = 62.4 lb/ft^3 = 0.0361 lb/in.^{3})

Solution : V_{billet} = (100 lb)/(2.7 x 62.4 lb/ft^3) = 0.594 ft^3 = 1026 in.^{3}

V_{bar} = 0.99(1026 in.^{3}) = (1.00-in. x 0.25-in. x L); L = 4060 in. = **338 ft.**

9-P13 A ceramic insulator will have a 1 v/o porosity after sintering and should have a length of 13.7 mm. During manufacturing, the raw powders can be compressed to contain 24 v/o porosity. What should the die dimension be?

Procedure : Relate both volumes, pressed and sintered to the true volume.

Solution : $V_{tr} = 0.76\ V_{pr} = 0.99\ V_{sint}$. $V_{pr}/V_{sint} = 0.99/0.76 = (L_{pr}/L_{sint})^3 = 1.303$.

$L_{pr} = L_{sint}\sqrt[3]{1.303} = 1.092(13.7\text{ mm})$ = **15.0 mm.**

9-P14 A powdered metal part has a porosity of 23 percent after compacting of the powders and before sintering. What linear shrinkage allowance should be made if the total porosity after sintering is expected to be 2 percent?

Basis : 1 mm^3 product = 0.98 mm^3 true volume + 0.02 mm^3 porosity.

Solution : Initial volume = 0.98 mm^3/0.77 = 1.27 mm^3 = $(1 + \Delta L/L_{sint})^3$.

$\Delta L/L_{sint}$ = **0.08.**

9-P15 A ceramic wall tile, 5 mm x 200 mm x 200 mm, absorbs 2.5 g of water. What is the apparent porosity of the tile?

Solution : 2.5 g H_2O = 2.5 cm^3 absorbed ⇒ 2.5 cm^3 pores per 200 cm^3 tile = **1.25 v/o** .

Comment : The *apparent* porosity involves open pores only. *True* porosity would also include closed pores

9-P16 A brick weighs 3.3 kg when dry; 3.45 kg when saturated with water, and 1.9 kg when suspended in water. (a) What is the apparent porosity? (b) What is its bulk density? (c) What is its ts apparent density?

Procedure : Refer to Example 9-1.2 Determine the open pore volume from absorption, and the bulk volume from bouyancy. Use cm^3.

Solution : Open pore volume = (3450 g − 3300g)/(1 g/cm^3) = 150 cm^3 .

Bulk volume = total H_2O displaced (bouyancy) = (3450 g −1900g)/(1 g/cm^3) = 1550 cm^3.

(a) Open (apparent) porosity = V_{open}/V_{bulk} = (150 cm^3)/(1550 cm^3) = 0.097 = **9.7 v/o.**

(b) Bulk density = m/V_{bulk} = (3300 g)/(1550 cm^3) = **2.13 g/cm^3** .

(c) Apparent density = $m/V_{app.}$ = (3300 g)/(1550 - 150 cm^3) = **2.36 g/cm^3** .

9-P21 Copper increases the strength of nickel, even though nickel is stronger than copper. Why?

Rationale : In nickel, as well as in any other crystalline solid, a solute atom will pin a dislocation, restraining movement (Fig 8-3.5).

9-P22 Which is stronger, a copper alloy with 2 w/o tin, or one with 2 w/o beryllium?

Procedure : Determine the a/o of the solutes, since the resistance to dislocation movements depend on the number of atoms. Use the data from Fig. 9-2.2(b).

Solution :	Atom %	Yield strength
98 amu Cu = 1.542 Cu		
2 amu Sn = 0.017 Sn	1.1	$1.3(S_y)_{Cu}$
2 amu Be = 0.222 Be	12.6	$\mathbf{3(S_y)_{Cu}}$ ←

9-P23 You are designing a seat brace for a motorboat. Iron is excluded because it rusts. Select the most appropriate alloy from Fig. 9-2.1. The requirements include an ultimate strength of at least 310 KPa (45,000 psi); a ductility of at least 45 percent elongation (in 50 mm); and low cost. (*Note* : Zinc is less expensive than copper, which in turn is less expensive than nickel.)

Sketch a "window" :

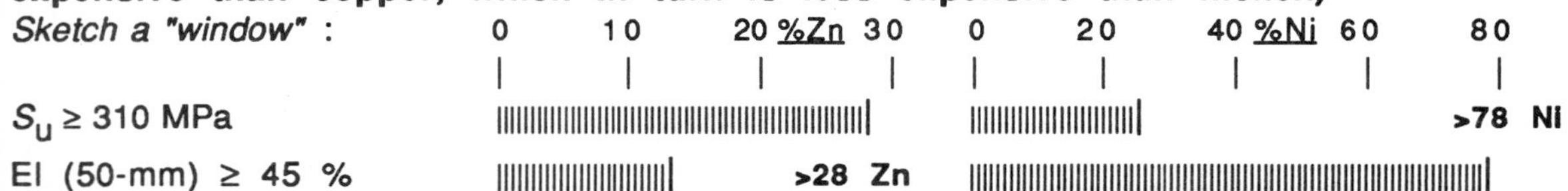

Either a brass with more than 28 % Zn, or a Cu-Ni alloy with more than 78 % Ni will meet the specifications. Since $\$_{Zn} < \$_{Cu} < \$_{Ni}$, use a brass with the maximum zinc. Normally, this is **65Cu-35Zn**, because of the solubility limit (Fig. 5-6.1).

Comment : Color displacement during printing may produce a slight variance in your answer.

9-P31 Two identical 20-m (65-ft) aluminum rods are each 14.0 mm (0.551 in.) in diameter. One of the two is drawn through a 12.7-mm (0.50-in.) die. (a) What are the new dimensions of that rod? (b) Assume test samples of identical size are machined from each rod (deformed and undeformed) and are marked with 50-mm gage lengths. Which one, if either, will have the greater ductility? Which one will have the greater yield strength?

Solution : (a) Since $V_1 = V_2$; $(20 \text{ m}/L_2) = (14 \text{ mm})^2/(12.7 \text{ mm})^2$. $L_2 =$ **24.3 m.**

(b) The **undeformed** bar will be **more ductile** . The deformed bar has already "used" some of its ductility before being prepared as a sample with gage marks.

The **deformed** bar will have higher **yield strength** because it contains more dislocations.

9-P32 (a) How much cold work was performed on the aluminum rod in Problem 9-P31? (b) Change the metal in Problem 9-P31 to copper. What are the ultimate strength and the ductility for the deformed metal?

Solution : (a) C.W. = $(A_o - A_1)/A_o$ = $(14.0^2 - 12.7^2)/(14.0^2)$ = 0.18, or **18 %** .

Alternate : Since $L_oA_o = L_1A_1$, C.W. = $(L_1 - L_o)/L_1$ = (24.3 − 20 m)/(24.3 m) = **18 %.**

(b) With C.W. = 18%, and using Fig. 9-3.4(b), S_u = **280 MPa**; and El.(50-mm) = **22%.**

9-P33 A copper wire 2.5 mm (0.10 in.) in diameter was annealed before being cold drawn through a 2.0-mm (0.08-in.) die. What ultimate strength does the wire have after cold drawing?

Solution : C.W. = $(d_o^2 - d_f^2)/d_o^2$ = (6.25 - 4.0 mm^2)/6.25 m^2 = 0.36, or 36 % .

With C.W. = 36%, and using Fig. 5-3.4(b), S_u = **330 MPa,** or 48,000 psi.

Comment : The elongation data in the figures of this chapter are for the standard 50-mm test bar (Fig. 8-2.2), in which the diameter is 0.505 in. (12.8 mm). Since the necking before breaking is a function of the diameter, ductility values for wires normally require a correction.

9-P34 Use the data of Figs. 8-3.6, 9-3.4, and 9-3.6. Sketch a plausible curve of yield strength vs. cold work for copper.

Sketch : Use the copper curve of Fig. 9-3.4(b) as a base. From Fig. 8-3.6, $S_y \approx$ 40 MPa (6000 psi) at 0% C.W. With C.W., the metal loses ductility so that S_y approaches S_u , as demonstrated for steel Fig. 9-3.6).

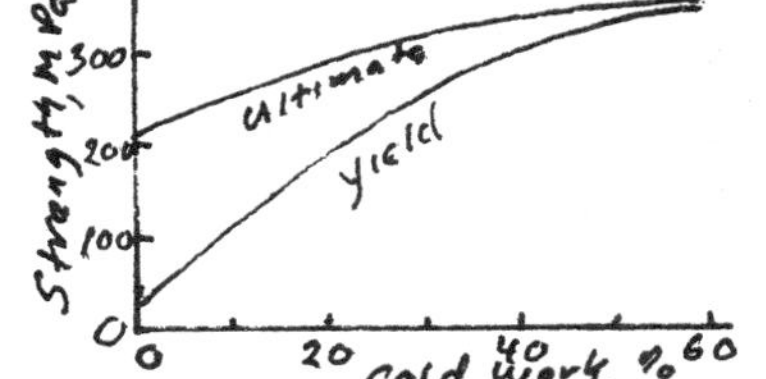

9-P35 A copper wire must have a diameter of 0.7 mm and an ultimate strength >345 MPa (50,000 psi) It is to be processed from a 10-mm rod. What should the diameter be for annealing before the final cold draw?

Procedure : For S_u = 50,000 psi (345 MPa), 45% C.W. is required (Fig. 9-3.4b). A one-step cold-draw would be 99.5 % C.W.!! Even if drawn into a wire without breaking, the product would have negible ductility. Therefore, we need an alternate process.

Solution : C.W. = 0.45 = $[d^2 - (0.7 \text{ mm})^2]/d^2$. d = **0.95 mm.**

Comment: Reduce the rod from 10 mm to 0.95 mm, either by hot work, or in steps by (cold work + anneal) cycles. Anneal at 0.95 mm, then cold work (cold draw) 45% from 0.95 to 0.7 mm diameter. (Also see Comment with the solution of Problem 9-P33.)

9-P36 A 70-30 brass wire is to be made by cold drawing with an ultimate strength of more than 415 MPa (60,000 psi), a hardness of less than 75 R_b, and an elongation of greater than 25 percent on a standard test bar. The diameter of the wire as received is 2.5 mm (0.10 in.). The diameter of the final product is to be 1.0 mm (0.04 in.). Prescribe a procedure for obtaining these specifications.

Sketch a "window" :

$S_u \geq$ 415 MPa

Hardness $\leq$ 75 R_b

El. (50-mm) $\geq$ 25 %

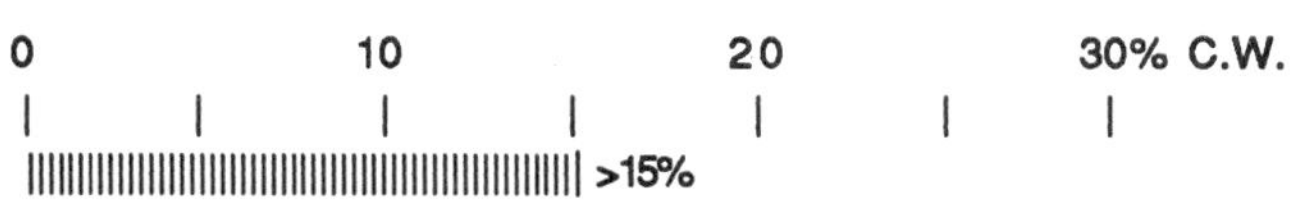

Window: 15-19 %. Use 17 %. 0.17 = $[d_o^2 - (1.0 \text{ mm})^2]/d_o^2$ d_o = 1.1 mm

Choice 1: **Hot work from 2.5 ⇒ 1.1 mm; cold work 17 % to 1.0 mm.**

Choice 2: **Cold work- anneal cycles ⇒1.1 mm; anneal; cold work 17 % ⇒1.0 mm.**

Comment : See the Comments with the solution of Problems 9-P23 and 9-P33.

9-P37 A rolled 66 Cu-34 Zn brass plate 12.7 mm (0.5 in.) thick has a ductility of 2% elongation by the standard test bar (Fig. 8-2.2) when it is received from the supplier.

This plate is to be rolled to a sheet with a final thickness of 3.2 mm (0.125 in.). In this final form, it is to have an ultimate strength of at least 483 MPa (70,000 psi) and a ductility of at least 7 percent elongation. Assume that the rolling process does not change the width.

Specify *all steps* (including temperatures, times, thickness, etc.) that are required.

Procedure : Since the plate is highly cold-worked as received (El. = 2%), it will miss all of the specifications. For cold work data , extrapolate from Fig 9-3.5.

Step 1 : Anneal above 350°C for 1 hr (based on Fig. 9-3.9) to obtain a starting point.

For $S_u \geq$ 70,000 psi, C.W. ≥ ~26 %. \
For El(2-in.) ≥7 %, C.W. ≤ ~45 %. / Use **30% C.W.**

$0.30 = (tw - 0.125w)/tw$ $\quad t$ = **0.18 in.** (or 4.6 mm).

Step 2 C.W. 64 % to 0.18 in. (4.6 mm) \
Step 3 Anneal >360°C, 1 hr. / or Hot Work
Step 4 Cold work 30 % to 0.18 in. (3.2 mm).

Comment : The comment with Problem 9-P33 also applies to thin sheet product..

9-P38 Aluminum has been shaped into a cake pan by spinning (Fig. 9-1.3b). Assume that the data from Fig. 9-3.12 apply to this cold worked aluminum. Will the metal start to recrystallize while in a 180°C (350°F) oven? Solve both (a) graphically, and (b) mathematically.

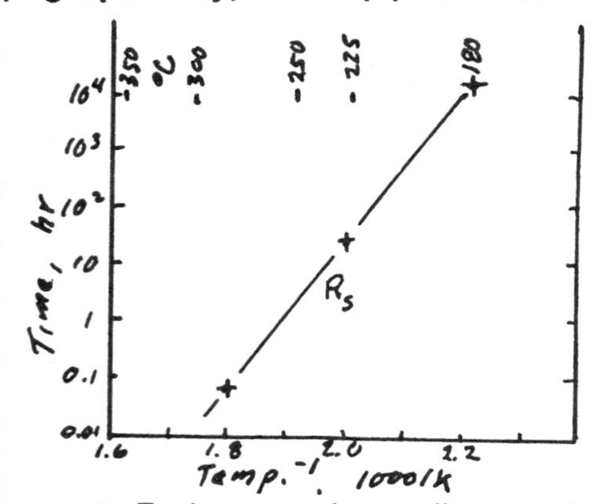

(a) At 180°C, 1000/K = 2.21
By extrapolation of the R_s curve,
$\log_{10} t \cong$ ~4.1; $\quad t$ =~ **12,500 hr.**

(b) At 1000/K =2.0, $t \cong$ 24 hr.
$\ln t = C + B/500\text{ K} = 3.2$
At 1000/K = 1.8, $t \cong$ 0.06 hr.
$\ln t = C + B/556\text{ K} = -2.8$
B = 29,800 K; C = -56.4

At T = 180° + 273° = 453 K,
$\ln t = -56.4 + 29800/453$; t = ~ **11,500 hr**

Comment: Each answer is equally accurate, since the points were taken from the same data.

9-P39 Assume that annealing should be completed in 1 sec. to permit the hot working of the aluminum in Fig. 9-3.12. What temperature is required?

Solution : Graphically :- 1000/K = 1.45 by extrapolation. $\quad T$ = 690 K ≅ **417°C.**

Alternate: Repeat the procedure of Problem 9-P38 and Example 9-3.4 to obtain:

B = 30,000 K, and C = -52

$\ln$ [1 hr/(3600 s/hr)] = -52 + 30,000/T ; $\quad T$ = 685 K ≅ **412°C.**

9-P41 Explain why a 92 percent copper, 8 percent nickel alloy can (or cannot) be age hardened.

Explanation : Refer to Fig. 5-2.2. The two metals form a continuous series of solid solutions. Therefore, supersaturation does not occur, and no precipitation can follow.

Age hardening is **not** possible without precipitation from a supersaturated solid solution.

9-P42 Maximum hardness is obtained in a metal when an aging time is 10 sec at 380°C, or 100 sec at 315°C. Neither of these constraints is satisfactory for production because we cannot be certain the parts will be uniformly heated. Recommend a temperature for the maximum hardness in 1000 sec (15 to 20 min), a time compatible with production requirements.

Procedure : Age hardening requires thermal activation for atom movements; therefore, the Arrhenius relationship enters. We can assume that $\ln t = C + B/T$ holds.

Solution :
$\ln 10 = C + B/(380 + 273\text{ K})$ \
$\ln 100 = C + B/(315 + 273\text{ K})$ / Solving, $C = -18.53$; $B = 13{,}600\text{ K}$.
$\ln 1000 = -18.53 + 13{,}600\text{ K}/T$; $T \approx 535\text{ K} \cong$ **260°C** (or 500°F)

9-P43 How long should it take the metal in the Problem 9-P42 to reach maximum hardness at 100°C?

Solution : Use the same Arrhenius-type equation.
$\ln t = -18.53 + 13{,}600\text{ K}/(373\text{ K})$; $t \approx 6\text{x}10^7\text{ s} \approx$ **2 yrs**

9-P44 An aircraft manufacturer receives a shipment of aluminum-alloy rivets that are already age hardened. Can they be salvaged? Explain your answer.

Procedure : Reheat into the solid solution range, i.e. *solution treat* , and cool to subambient temperatures. The temperature and time depend on the alloy, e.g. 96 Al-4 Cu amd 95 Al-5 Mg, among others.

9-P51 More rapid cooling produces a pearlite with twice as many layers as the one shown in Fig. 9-5.2(a). How much (what percentage) change will there be in the amount of interphase boundary that is available for plastic constraint?

Procedure : The amount of interphase boundary area can be determined by the intercepts along a random line (Example 7-1.1). With twice as many layers, P_L will double, as will S_V of Eq. (7-1.2). Therefore, **100% increase** of interphase boundary area.

9-P52 Extended annealing produces a spheroidite (Fig. 9-5.2b) with carbide particles that are twice as large in diameter. (See $\overline{L}$ of Eq. 7-1.1.) How much (what percentage) change will there be in the amount of interphase boundary that is available for plastic constraint?

Answer : The $\overline{L}$ across the carbides will be twice as long. Therefore, P_L and S_V will be half as large. Therefore, **50% decrease** of interphase boundary area.

9-P61 (a) Determine the temperature that should be used to normalize (a) a 1030 steel, (b) a 1080 steel, and (c) a 1% carbon steel.

Procedure : Normalizing requires that the steel be heated approximately 50°C (~100°F) into the austenite range. Check the lowest temperature for complete austenite in Fig. 7-3.1. The exact temperature will depend on the heating time (size of steel item).

Solution: (a) $\gamma_{0.3}$ = 800° + ~50°C ≅ **850°C,** or $\gamma_{0.3}$ = 1475° + ~100°F ≅**1575°F**

(b) $\gamma_{0.8}$ = 727° + ~ 50°C ≅ **780°C,** or $\gamma_{0.8}$ = 1340° + ~100°F ≅**1440°F**

(c) $\gamma_{1.0}$ = 805° + ~ 50°C ≅ **855°C,** or $\gamma_{1.0}$ = 1480° + ~100°F ≅**1580°F**

9-P62 A small piece of 1080 steel is heated to 800°C, quenched to -60°C, reheated immediately to 300°C, and held 10 sec. What phases are present at the end of this time?

Procedure : This involves an γ—>M reaction before ($\alpha + \overline{C}$) can form. There is no composition change (Fig. 9-6.4 for 1080 steel).

Answer : At **800°C**: all austenite. At **-60°C**: all martensite.

At **300°C** plus 10 s: martensite. It will evenutally change to ($\alpha + \overline{C}$); however, very little progress in 10 seconds.

Comment : It will not go back to austenite unless the temperature is raised above the eutectoid temperature where γ is stable. Equation (7-4.1) is irreversible below 727°C.

9-P63 A piece of 1045 steel is quickly quenched from 850°C to 400°C and held (a) for 1 sec, (b) for 10 sec, (c) for 100 sec. What phases(s) will be present at each time point?

Answers: Fig. 11-6.3. (a) 1 s: **only γ.** (b) 10 s: γ **+** (α **+** $\overline{C}$). (c) 100 s: (α **+** $\overline{C}$)

9-P64 Sketch a plausible isothermal transformation diagram for a steel of 1.0 w/o C and 99.0 w/o Fe.

Comment :The right side of the I-T diagram must match the phase diagram. Note that added carbon lowers the M_s and M_f temperatures (Figs. 9-6.3 and 9-6.4). Also, reactions are slowest when γ is stable at lower temperatures (eutectoid).

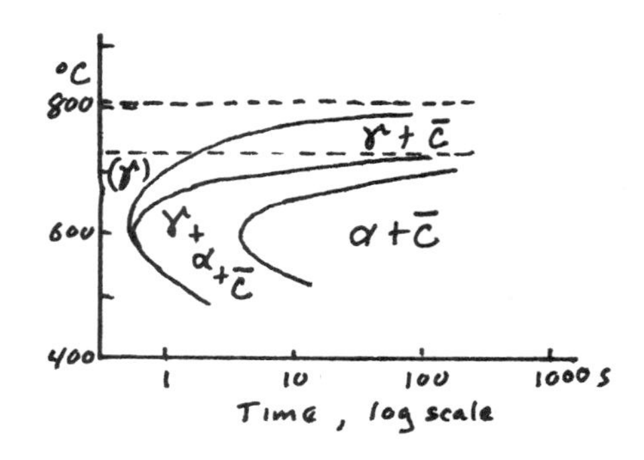

9-P65 A small piece of 1080 steel has its quench interrupted for 20 sec at 300°C (570°F) before final cooling to 20°C. (a) What phase(s) are present after the 20 sec? (b) What phase(s) are present at 20°C?

Answer : Assume prior austenization. (a) At 300°C, (0 sec): γ. At 300°C, (20 sec): still γ.

(b) At 20°C: **M + some retained γ.**

Comment: Both the γ and the M contain 0.8 % carbon.

9-P66 Why did we *not* use Fig. 9-6.3 for wire (c), step 2, of Example 9-6.1?
Explanation : At 730°C, at 1045 steel contains a mixture of α and γ, the latter containing ~0.8 % carbon. Thus, it decomposes identically to γ in 1080 steel (Fig. 9-6.4). The α doesn't change when cooled.

9-P67 A 1020 steel is equilibrated at 770°C (1420°F), then is quenched rapidly to 400°C (750°F). How long must the steel be held at that temperature to reach the midpoint of the austenite decomposition?
Analysis : At 770°C, a 1020 steel contains α(nil C), + γ(0.45 % C). Quenching does not affect the ferrite. The γ is the same as that found in autentized 1045 steel. Therefore, use Fig 9-6.3. At 400°C, $t_{50\%} \cong$ **8 s.**

9-P68 Compare and contrast the following terms: (a) *martensite* and *tempered martensite*; (b) *tempered martensite* and *spheroidite*; (c) *tempered martensite* and *pearlite*.
(a) M. - Bct phase of iron, unstable, supersaturated with carbon
T.M.- A microstructure of carbide particles in a ferrite matrix. Comes from the decomposition of martensite
(b) T.M. and Sph. - Both possess a microstructure of $\overline{C}$ particles in a α matrix.
Sph. - The carbide particles are larger than in T.M. Obtained by either over-tempering martensite, or by extended annealing of pearlite.
(c) T.M. and P.- Both microstructures of ($\alpha + \overline{C}$).
P. - lamellar $\overline{C}$ from γ decomposition. T.M. - particulate $\overline{C}$ from M. decomposition.

9-P69 Why does austenite in a 0.5% Mo-0.7% C steel transform to ($\alpha + \overline{C}$) less rapidly than it does in an 0.7 % C steel?
Rationale : $R_{Mo} \approx R_{Fe}$; $r_C \ll R_{Fe}$. Molybdenum is substitutional in iron phases; carbon is interstitial. The moly must partition with the carbon, but diffuses more slowly . Thus it holds up the formation of (α + $\overline{C}$) .

9-P71 Explain the difference between *hardness* and *hardenability*.
Hardness: Resistance to penetration (or to scratching, if there is movement).
Hardenability: "Ability" to produce maximum hardness.

9-P72 (a) What is the cooling rate (reported in °C/sec) at the midradius of a 50-mm (1.97-in.) round steel bar, which was quenched in agitated oil? (b) What is it when reported as the distance from the quenched end of a Jominy bar? (c) Repeat parts (a and b), but for the 3/4-radius and a water quench.
(a) From upper abscissa of Fig. 9-7.5(b),~**17°C/s**. (b) From lower abscissa, D_{qe} = **13 mm.**
(c) From Fig. 9-7.5(a), ~**50°C/s**; and $D_{qe} \cong$ **5 mm.**
Comment : These are the cooling rates at the standard reference temperature of 700°C. Of course, the rate decreases as the temperature falls.

9-P73 Why should the steel (or the water) be agitated when a gear is water quenched?

Explanation : Steam forms at the surface of the hot steel. If a steam bubble remains attached to the surface, that area will be insulated, and subsequent heat extraction will be slow. This permits more ($\alpha + \overline{C}$) and less of the hard martensite. Agitation removes any attached bubbles.

Comment : A salt brine is sometimes used for the fastest possible quench, because steam bubbles release more readily from the surface than with water alone.

9-P74 (a) What is the quenched hardness at the midradius of a 50-mm (1.97-in.) round steel bar of an SAE 1040 steel quenched in agitated oil? (b) What is it when the steel is quenched in agitated water?

Answer : (a) Oil: From Fig. 9-7.5(b), D_{qe} = 13 mm; from Fig. 9-7.4, **25 R_C**.

(b) Water: From Fig. 9-7.5(a), D_{qe} = 7.5 mm; from Fig. 9-7.4, **28 R_C**.

9-P75 (a) What hardness would you expect at the center of a 50-mm (2-in.) round bar of an SAE 1040 steel if that bar were quenched in agitated oil? (b) What hardness if it were quenched in agitated water?

Answer : (a) Oil: From Fig. 9-7.5(b), D_{qe} = 16.5 mm; from Fig. 9-7.4, **24 R_C**.

(b) Water: From Fig. 9-7.5(a), D_{qe} = 10.5 mm; from Fig. 9-7.4, **26 R_C**.

9-P76 A 40-mm (1.6-in.) bar of an SAE 1040 steel (i.e., with diameter = 40 mm, and length »40 mm) is quenched in agitated water.

(a) What is the cooling rate through 700°C at the surface?

(b) What is it at the center?

(c) Plot a hardness traverse.

Procedure : Obtain the cooling rate (Fig. 9-7.5).

From this, you may determine the hardness (Fig.9-7.4).

Answer : (a) Surface: D_{qe} = **1.5 mm; ~200°/s.**

Center: D_{qe} = **9 mm; ~30°/s.**

(b)

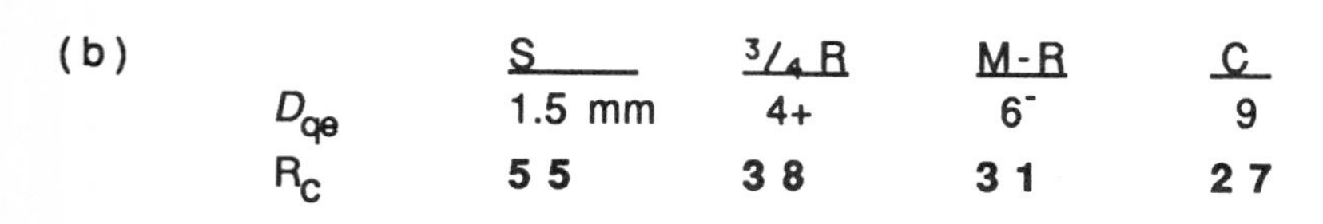

	S	¾ R	M-R	C
D_{qe}	1.5 mm	4+	6⁻	9
R_C	**55**	**38**	**31**	**27**

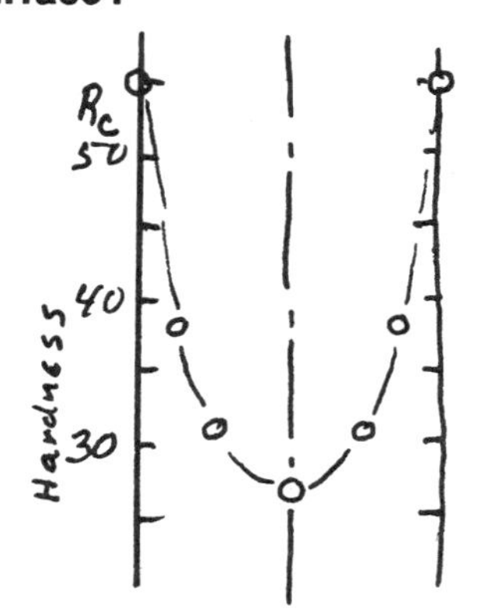

9-P77 Explain why low-alloy steels rather than plain-carbon steels are specified for components that are to be heat treated by quenching and tempering.

Explanation : Alloying elements slow down the $\gamma \longrightarrow (\alpha + \overline{C})$ reaction as shown in Fig 9-6.5. All of the metallic solutes (Mo, Ni, W, Cr, Si, Ti, V, and so on) diffuse more slowly than does carbon. These metallic elements must partition between the ferrite and carbide. This takes time. As a result, one may quench larger parts, or cool a part at a slower rate, and avoid ($\alpha + \overline{C}$) to obtain the harder martensite.

9-P78 An 80-mm (3.15-in.) round SAE 4340 steel bar is quenched in agitated oil. Plot the hardness traverse.

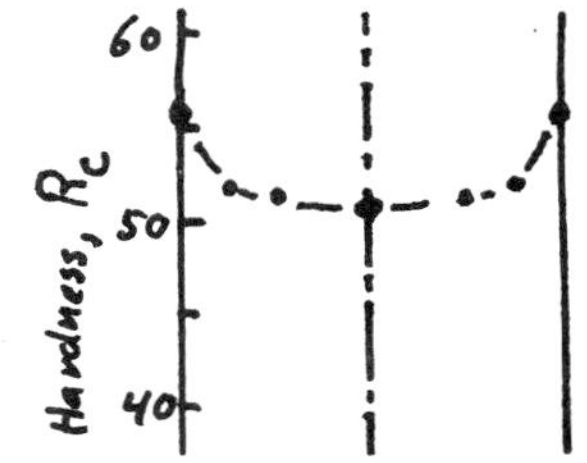

		S	3/4 R	M-R	C
Fig.9-6.5	D_{qe}	11 mm	18.5	22	27
Fig.9-6.4	R_c	56	53	52	51

9-P79 Repeat Example 9-7.5, but with oil quenching.

Location	D_{qe}	(a) Carburized		(b) Non-carburized	
Surface	6.5 mm	0.62% C	34 R_c	0.20 % C	23⁻ R_c
2-mm	7	0.35	28	"	
3/4 radius	7.5	0.20	22+	"	22+
M-R	8.5	"	22	"	22
Center	10	"	21	"	21

9-P81 Glass-coated (porcelain-enameled) steel is used as an oven liner.
(a) Why is this combination chosen?
(b) How should the properties of the glass and of the steel match? How should they be unlike?

Analysis : (a) Glass gives oxidation protection. Glass is easier to clean.
(b) They should possess a strong interfacial bond. This is established when the glass is molten. Their thermal expansion coefficients should be similar, but not identical. α_{gl} should be slightly less than α_{Fe}. In this manner, the glass will be placed under compression (Fig. 9-8.2) as the oven is cooled from above T_g , the glass softening temperature. (All kitchen use is below this temperature.)

9-P82 A 1-mm-diameter fiber with the composition of plate glass (Appendix C) is coated with 0.1 mm of borosilicate glass, so the fiber is now 1.2 mm in diameter.

Assuming no initial stresses at 200°C, what longitudinal stresses are developed when the composite fiber is cooled to 20°C?

Analysis : $F_p + F_b = 0$. $(\Delta L/L)_p = (\Delta L/L)_b$. Use Eq. (9-8.1). Compression = minus.

Solution : $s = F_p/A_p = F_p/[(\pi/4)(10^{-3}\text{m})^2] = 1.27 \times 10^6(F_p)$ Pa .

$s_b = -F_p/A_b = -F_p/[(\pi/4)(1.2 \times 10^{-3}\text{m})^2 - (\pi/4)(10^{-3}\text{m})^2] = -2.89 \times 10^6(F_p)$ Pa .

$$(9 \times 10^{-6}/°\text{C})(-180°\text{C}) + [(1.27\, F_p \text{ MPa})/70{,}000 \text{ MPa}]$$
$$= (2.7 \times 10^{-6}/°\text{C})(-180°\text{C}) + [(-2.89\, F_p \text{ MPa})/70{,}000 \text{ MPa}]$$

Solving, $F_p \approx 19$ N; $s_p \approx$ **+25 MPa,** or +3500 psi; $s_b \approx$ **-55 MPa,** or -8000 psi

Comment : The **surface** (borosilicate) is in **compression** (-); **strengthening** results.

9-P83 Repeat the Problem 9-P82, but interchange the two locations of the glasses. (a) Which will have the higher stress, plate or glass? (b) Comment on the strength of this composite glass rod.

Interchange A_p and A_b Thus, $s_p = 2.89 \times 10^6 (F_p)$ Pa, and $s_b = -1.27 \times 10^6 (F_p)$ Pa.

$$(9 \times 10^{-6}/°C)(-180°C) + [(2.89\ F_p\ \text{MPa})/70{,}000\ \text{MPa}]$$

$$= (2.7 \times 10^{-6}/°C)(-180°C) + [(-1.27\ F_p\ \text{MPa})/70{,}000\ \text{MPa}]$$

Solving, $F_p \approx 19$ N; $s_p \approx$ **+55 MPa,** or +8000 psi; $s_b \approx$ **-25 MPa,** or -3500 psi.

Comment : The **surface** (plate) is in **tension** (+); therefore, **cracking** occurs readily.

10-P11 Estimate the viscosity of polyvinyl chloride at 105°C.

Estimate : T_g of PVC is 85°C (Table 4-2.2). Therefore, using Eq (10-1.4a) ΔT = 20°C :

$\log_{10} \eta_{105°C} \approx 12 - (17.5 \times 20°C)/(52 + 20°C) = 7.14$; $\eta_{105°C} \approx$ **14 x10^6 Pa·s.**

10-P12 The viscosity of water is 0.001 Pa·s at 20°C, and 0.00028 Pa·s at 100°C. (a) What is the activation energy for flow? (b) What is the viscosity at 0°C?

Solution :(a) $\ln 0.001 = \ln \eta_o + E/k(293\ K) = -6.9$ \ Solving, E = **24 x10^{-21}J**;

$\ln 0.00028 = \ln \eta_o + E/k(373\ K) = -8.2$ / $\ln \eta_o = -12.8$.

(b) $\ln \eta_{0°C} = -12.8 + [24 \times 10^{-21} J/(13.8 \times 10^{-24} J/K)(273\ K)]$; $\eta_{0°C}$ = **0.0016 Pa·s**

10-P21 Forty-five kg of silica flour (finely ground quartz sand) are mixed with each 100 kg of melamine-formaldehyde (mf). What is the volume fraction of filler?

Data : From the Appendix. *Basis* : 100 g mf + 45 g SiO_2 .

Mf: 100 g/(1.5 g/cm^3) = 67 cm^3 = 0.80

SiO_2: 45 g/(2.65 g/cm^3) = 17 cm^3 = **0.20** , or 20 v/o .

10-P22 A 163-cm (64-in.) diameter outdoor sign is vacuum-formed into its final shape. What force would be required behind a die to provide the same force in a die-stamping operation as was obtainable by a vacuum?

Procedure : Assume essentially full vacuum of −14.7 lb$_f$/in.^{2} (− 100,000 Pa).

Solution : (− 14.7 psi)(π/4)(64 in.)^2 = − **47,300 lb$_f$** (compression)

or (− 100,000 Pa)(π/4)(1.63 m)^2 = − 210,000 N.

10-P41 A steel-reinforced aluminum wire has the same dimensions as does the steel-reinforced copper wire of Example 10-4.2. What fraction of the load will be carried by each metal?

Procedure : The strains must be equal; therefore, $e_{Al} = e_{St} = [(F/A)/E]_{Al} = [(F/A)/E]_{St}$

Solution : $F_{St}/F_{Al} = [E_{St}A_{St}]/[E_{Al}A_{Al}]$ = [205 GPa/70 GPa][0.8/2.4] = 0.98;

$F_{St}/(F_{St} + F_{Al}) = 0.98 F_{Al} / 1.98 F_{Al}$ =**0.49;** $F_{Al}/(F_{St} + F_{Al})$ = **0.51.**

10-P42 Derive a mixture rule for the conductivity (electrical or thermal) of a laminate when the heat flow is parallel to the structure.

Background : From Physics, for parallel resistances, $1/R = 1/R_1 + 1/R_2$.

Derivation : $A/\rho L = A_1/\rho_1 L + A_2/\rho_2 L$. Thus, $\sigma_m = f_1\sigma_1 + f_2\sigma_2 = \Sigma f_i\sigma_i$,

where f = vol. fract. Also, for thermal conductivity, $k_m = f_1k_1 + f_2k_2 = \Sigma f_ik_i$.

10-P43 Calculate the density of a glass-reinforced plastic fishing rod, in which the glass-fiber content is 15 w/o. (A borosilicate glass is used for the longitudinal fibers. The density of the plastic is 1.3 g/cm^3.)

Basis : 100 g = 15 g glass + 85 g plastic. Data from App.C.

Solution : Plastic: 85 g/(1.3 g/cm^3) = 65.4 cm^3 \ ρ = 100 g/(71.65cm^3)

Glass: 15 g/(2.4 g/cm^3) = 6.25 cm^3 / = **1.40 g/cm^3**.

10-P44 Estimate the thermal conductivity (longitudinal) of the composite in Problem 10-P43. (k_{pl} = 0.00026 W/mm·°C.)

Procedure: Refer to Prob. 10-P42, where parallel conductivity was considered. Data: App. C.

Solution: f_{pl} = 65.4 cm³/71.85 cm³ = 0.91. f_{gl} = 6.25 cm³/71.85 cm³ = 0.09 .

$k_m = f_1k_1 + f_2k_2$ = 0.91(0.00026) +0.09(0.0010 W/mm·°C) = **0.00033 W/mm·°C** .

10-P45 An AISI-SAE 1040 steel wire (cross-section 1 mm²) has an aluminum coating such that the total cross-sectional area is 1.2 mm². (a) What fraction of a 450 N load (100 lb$_f$) will be carried by the steel? (b) What is the resistance of this composite wire per unit length?

Procedure: Each metal will have the same strain: $e_{1040} = e_{Al} = e = (F/A)/E$). Data: App.C.

Solution: (a) $(F_{1040}/1\text{ mm}^2)/205\text{ GPa} = (F_{Al}/0.2\text{ mm}^2)/70\text{ GPa}$.

$F_{1040}/F_{Al} \approx 15$; thus, the steel carries **15/16** of the load, or 94%.

(b) $R = (R_1^{-1} + R_2^{-1})^{-1} = [(A/\rho L)_{1040} + (A/\rho L)_{Al}]^{-1}$

$= [10^{-6}\text{ m}^2/171\times10^{-9}\Omega\cdot\text{m} + 0.2\times10^{-6}\text{ m}^2/29\times10^{-9}\Omega\cdot\text{m}]^{-1}$ = **0.08 ohm/m** .

10-P46 A glass-reinforced plastic rod (fishing pole) is made of 67 v/o borosilicate glass in a nylon matrix. What is the longitudinal thermal expansion coefficient?

Procedure: $(\Delta L/L)_{gl} = (\Delta L/L)_{ny}$. Since $F_{gl} = -F_{ny}$, and $A_{gl} = 2A_{ny}$, $s_{ny} = -2s_{gl}$.

Solution (+1°C): $\Delta L/L = \alpha\Delta T + sE$ = $(2.7\times10^{-6}/°C)(+1°C) + s_{gl}/70{,}000$ MPa

= $(100\times10^{-6}/°C)(+1°C) + (-2s_{gl})/2{,}800$ MPa.

Therefore, s_{gl} =133,500 Pa, (and s_{ny} = -0.267 MPa).

Substituting: $(\Delta L/L)_{+1°} = \alpha_{comp.}$

= $(2.7\times10^{-6}/°C)(+1°C)$ + 0.1335 MPa/70,000 MPa = **4.6 x10⁻⁶/°C**

= $(100\times10^{-6}/°C)(+1°C)$ + (-0.267 MPa)/2,800 MPa = **4.6 x10⁻⁶/°C.**

10-P51 What is the mass of a mer of the polycellulose molecule?

From Fig. 10-5.1, $C_6H_7(OH)_3O_2$; therefore, 72 + 7 + 51 + 32 amu = **162 amu** .

10-P52 Birchwood veneer is impregnated with phenol-formaldehyde (PF) to ensure resistance to water and to increase the hardness and dimensional stability of the final product. Although this wood weighs only 0.56 g/cm³, the true density of th7 cellulose-predominant material is 1.52 g/cm³. (a) How many grams of PF are required to impregnate 10,000 mm³ (0.6 in.³) of dry birchwood? (b) What is the final density?

Solution: 1 cm³ basis.

Pore vol. = [1 cm³ - (0.56 g/1.52 g/ cm³)] = 0.632 cm³. From App. C, ρ_{PF} = 1.3 g/cm³.

(a) (0.632 cm³)(1.3 g/ cm³) = 0.82 g PF per cm³ of wood, or **8.2 g**/10,000 mm³.

(b) (0.82 g PF + 0.56 g wood)/cm³ = **1.38 g/cm³.**

11-P11 **(I^2R review) A dry cell (1.5 V) is connected across a 10-ohm circuit. (a) What is the resulting current?) (b) What is the wattage or energy rate? (c) How much energy is used per hour?**

Recall from Physics : $E = IR$, and $P = EI = E^2/R$

(a) I = 1.5 V/10 ohm = **0.15 amp**

(b) P = (1.5 V)(0.15 A) = 0.225 V·A = **0.225 J/s**; or P = (1.5 A)^2/10 Ω = **0.225 W.**

(c) E = (0.225 J/s)(3600 s) = **810 J**; or E = (0.225 W)(1 h) = **0.225 Wh.**

11-P12 **(I^2R review) An 18-m copper wire connects two terminals of a dry cell (1.5 V). (a) What is the current density in the wire? (b) How many watts go through the wire, if it has a 3-mm^2 cross-section? (c) What diameter must be used to limit the current to 1 amp?**

Background : Current density, J of Eq. (1-2.2), may also be identified as i.

From Eq. (1-2.2), $i = \sigma \mathcal{E} = E/L\rho$. (We'll use $\mathcal{E}$ for electric field, V/m.)

Solution : (a) i = 1.5 V/(17 x10^{-9} ohm·m)(18 m) = **5 x10^6 A/m^2.**

(b) $P = EI = EiA$ = 1.5 V(5 x10^6 A/m^2)(3 x10^{-6} m^2) = **22.5 W**

(c) Rearranging, $A = I\rho L/E$ = (1 A)(17x10^{-9} ohm·m)(18 m)/1.5 V = 0.20 x10^{-6} m^2.
= 0.20 mm^2. d = **0.5 mm .**

11-P13 A wire must have a diameter of less than 1 mm, and a resistance of less than 0.1 ohm/m. Which of the materials in Appendix C are suitable?

Solution : $R/L = \rho/A \leq 0.1$ ohm/m.

ρ = (0.1 ohm/m)(π/4)(<0.001 m)^2 ≤ 78 x10^{-9} ohm·m .

Therefore, **Al, Al alloys, brass, Cu, Mg, sterling silver.**

11-P14 Seventy mV are placed across the 0.5-mm dimension of a semiconductor with the carrier mobility of 0.23 m^2/v·s. What drift velocity develops?

Solution : $\mu = \bar{v}/\mathcal{E}$; $\bar{v}$ = (0.23 m^2/V·s)(0.070 V/0.0005 m) = **32 m/s** .

Comment : This is a drift,or net, velocity. Of course, the path includes many deflections.

11-P15 Charge carriers within a semiconductor, which has the resistivity of 0.0313 ohm·m, possess a drift velocity of 6.7 m/s when a 1.2 V differential is placed across a 9-mm piece. How many carriers are there per m^3?

Procedure : We can calculate the mobility, from which we can get the carrier concentration.

Solution : μ = (6.7 m/s)/(1.2 V/0.009 m) = 0.050 m^2/V·s .

n = (1/ρ)/μq = (1/0.0313 ohm·m)/(0.050 m^2/V·s)(0.16 x10^{-18}A·s) = **4 x10^{21}/m^2.**

11-P16 Laboratory measurements indicate that the drift velocity of electrons in a semiconductor is 149 m/s when the voltage gradient is 15 V/mm. The resistivity is 0.7 ohm·m. What is the carrier concentration?

Procedure : From Eqs. (11-1.1 and 11-1.2), $n = \sigma/q\mu = \mathcal{E}/\rho q\bar{v}$.

Solution : n = [15,000 V/m]/[0.7 ohm·m)(0.16 x10^{-18}A·s)(149 m/s)] = **9 x10^{20}/m^3**

11-P17 List and compare the various types of charge carriers in solids.

Types:		
	Electrons-	Negative, delocalized.
	Anions-	Atoms with excess electrons, negative.
	Cations-	Atoms with missing electrons, positive.
	Electron holes-	Positive, missing delocalized electrons.

11-P21 Based on Fig. 11-2.2 what is the electrical resistivity of an annealed 70-30 brass?

Solution: $\sigma_{70\text{-}30} = 0.28\,\sigma_{Cu}$; $\rho = 1/[0.28(60\times10^{6}\ \text{ohm}^{-1}\cdot\text{m}^{-1})] =$ **60 ohm·nm**

Alternate: From App.C, $\rho_{Cu} = 17\times10^{-9}\ \Omega\cdot\text{m}$; $\rho_{70\text{-}30} = 17\ \Omega\cdot\text{nm}/0.28 =$ **60 ohm·nm**

11-P22 Estimate the resistivity of a tungsten wire at 1500°C.

Estimate: $(50\times10^{-9}\ \text{ohm}\cdot\text{m})[1 + (0.0045/°\text{C})(1500°\text{C})] =$ **390×10^{-9} ohm·m** .

11-P23 A 6% variation (maximum) is permitted in the resistance between 0° and 25°C. Which metals of Table 11-2.1 meet this specification?

Limits : $1.06 \geq [1 + y_T(25°\text{C})]$; therefore, $y_T \leq 0.0024/°\text{C}$.

Choices: Only the **alloys** of Table 11-2.1, none of the pure metals. This is because the solute resistivity, which is constant with temperature, commonly predominates over the thermal resistivity.

11-P24 Based on data of Table 11-2.1 and of Appendix C, (a) estimate the effect of copper on the resistivity of silver (ohm·nm per a/o). (b) Why do the data of Table 11-2.2 not apply?

Procedure : Here, as with many data sources, different bases are used. In App.C, 20°C is used because that is close to ambient. Table 11-2.2 uses 0°C because that permits °C to be used directly, rather than ΔT.

From App.C, $\rho_{Ster.Ag} = 18$ ohm·nm @ 20°C.

Solution: (a) $\rho_{pure\ Ag} = 15\ \text{ohm}\cdot\text{nm}\,[1 + 0.0038/°\text{C}(20°\text{C})] = 16.1$ ohm·nm @ 20°C.

From Fig. 5-2.5, 7.5 w/o = 12 a/o. (Or, this may be calculated from the atomic masses.)

$(18 - 16.1\ \text{ohm}\cdot\text{nm})/(12\ \text{a/o}) =$ **0.16 ohm·nm/(a/o Cu).**

(b) In Table 11-2.2, Cu is the solvent; Ag is the solute. In sterling silver, Cu is the solute.

11-P25 Using the data of Appendix C, determine the ratios of thermal to electrical conductivities, k/σ, for each metal. (b) Based on your calculations, make a generalization about the relationship between electrical and thermal conductivities.

Background: W-F ratio $= k/\sigma = k\rho$. Units = $[(\text{W/m}^2)/(°\text{C/m})](\text{ohm}\cdot\text{m}) = \text{W}\cdot\text{ohm/m}$.

Pure metals:		**Alloys :**	
Al:	$(220)(29\times10^{-9}) = 6.4\times10^{-6}$ W·ohm/m.	Al alloys:	$(160)(45\times10^{-9}) = 7.2\times10^{-6}$ W·ohm/m
Cu:	$(400)(17\times10^{-9}) = 6.8\times10^{-6}$ " .	Brass:	$(120)(62\times10^{-9}) = 7.4\times10^{-6}$ "
Fe:	$(72)(98\times10^{-9}) = 7.1\times10^{-6}$ " .	Bronze:	$(80)(135\times10^{-9}) = 11\times10^{-6}$ "
Pb:	$(33)(206\times10^{-9}) = 6.8\times10^{-6}$ " .	Ster.Ag:	$(410)(18\times10^{-9}) = 7.4\times10^{-6}$ "
Mg:	$(160)(45\times10^{-9}) = 7.2\times10^{-6}$ " .	Stain.St.:	$(15)(700\times10^{-9}) = 11\times10^{-6}$ "

11-P25 (con't) *Comment* : Pure metals and low alloys (< 10 a/o) generally have a W-F ratio of approximately 7×10^{-6} W·ohm/m . This provides a good procedure to estimate thermal conductivities. However, note that higher alloy contents exceed this k/σ ratio, e.g., stainless steel and monel metal.

11-P26 Estimate the thermal conductivity of copper at 40°C from its electrical conductivity.

Procedure : Calculate the electrical resistivity; then use the W-F ratio.

$$\rho_{40°C} = (16 \times 10^{-9}\ \text{ohm·m})[1 + 0.0039/°C(40°C)] = 18.5 \times 10^{-9}\ \text{ohm·m}$$

$$k_{40°C} = \text{W-F}/\rho = (7 \times 10^{-6}\ \text{W·ohm/m})/(18.5 \times 10^{-9}\ \text{ohm·m}) = \mathbf{380\ W·ohm/m}$$

11-P27 A brass must have a resistivity of ≤50 ohm·nm, and its hardness must be ≥50 R_F. (a) What range of zinc content is permissible? (b) What is the *best* choice for a composition within that range?

Sketch a "window" : 0 10 20 30% Zn.

(a)

Hardness ≥50 R_f (Fig. 9-2.1) |||||||||||||||||||||||||||||| >9%

$\rho \leq 50$ ohm·nm $= \sigma \geq 20 \times 10^6\ \text{ohm}^{-1}\cdot\text{nm}^{-1}$ (Fig.11-2.2) <22% ||||||||||||||||||||||||||||||||||||||/

Window: **9↔22 %.** (b)**Use 20 %**; cheaper, but still some working margin from upper limit.

11-P28 A brass wire must carry a load of 45 N (10 lb) without yielding and must have a resistance of less than 0.033 ohm/m (0.01 ohm/ft).
(a) Calculate the diameter of the smallest wire that can be used, if the wire is made of 60-40 brass, (b) 80-20 brass, (c) 100% Cu.

		60Cu-40Zn	80Cu-20Zn	100 Cu
S_y (Fig. 9-2.1)	MPa	80	70	50
ρ (Fig. 11-2.2)	ohm·m	70×10^{-9}	45×10^{-9}	17×10^{-9}
For S_y, $d > \sqrt{4F/(\pi S_y)}$	mm	0.8	0.9	**1.07 mm**
For ρ, $d > \sqrt{4\rho L/(\pi R)}$	mm	1.64 mm	1.32 mm	0.81
Use:(round diameters upward)		1.7 mm (0.07 in.)	1.4 mm (0.055 in.)	**1.1 mm** (0.042 in.)

11-P29 A certain application requires a piece of metal that has a yield strength greater than 100 MPa and a thermal conductivity greater than 0.04 W/m·°C. Specify either an annealed brass or an annealed Cu-Ni alloy that will meet the requirements.

Sketch a "window" : 0 10 20 %Zn 30 | 0 20 40 %Ni 60 80

$S_y \geq 100$ MPa ||||||||||||||||||||||||||||||||||||||||||||x ||||||||| >12% >80%↓

$k \geq 0.04$ W/m·°C <30% ||||||||||||||||||||||||||||||||||||||

"Window:" Brass: none. Cupronickel: 12 to 30%, and >80% Ni. Use **85Cu-15Ni.**

11-P31 Calculate the fraction of the energy states in which metal are occupied at $E = E_f + 0.05$ eV (a) at 100°C, (b) at 200°C, (c) at 400°C, and (d) at 800°C?

Procedure : Use the Fermi distribution Eq. (11-3.2).

(a) $F(E)_{100°C} = [1 + \exp(0.05\ \text{ev})/(86.1 \times 10^{-6}\ \text{ev/K})(373\text{K})]^{-1} = 1/(1 + e^{1.56}) = \mathbf{0.17}$

(b) $F(E)_{200°C} = [1 + \exp(0.05\ \text{ev})/(86.1 \times 10^{-6}\ \text{ev/K})(473\text{K})]^{-1} = 1/(1 + e^{1.23}) = \mathbf{0.23}$

(c) $F(E)_{400°C} = [1 + \exp(0.05\ \text{ev})/(86.1 \times 10^{-6}\ \text{ev/K})(673\text{K})]^{-1} = 1/(1 + e^{0.86}) = \mathbf{0.30}$

(d) $F(E)_{800°C} = [1 + \exp(0.05\ \text{ev})/(86.1 \times 10^{-6}\ \text{ev/K})(1073\text{K})]^{-1} = 1/(1 + e^{0.54}) = \mathbf{0.37}$

11-P32 (a) Calculate and plot the Fermi distribution for 20°C in increments of 0.05 eV from $E = E_f - 0.15$ eV to $E = E_f + 0.15$ eV. (b) Explain what this distribution means. $[F(E) = 1/(1 + e^{[E - E_f]/kT})]$

$E_f + 0.15$ eV $\quad F(E)_{293K} = 1/(1 + e^{5.95}) = 0.003$

$E_f + 0.10$ eV $\quad = 1/(1 + e^{3.96}) = 0.019$

$E_f + 0.05$ eV $\quad = 1/(1 + e^{1.98}) = 0.121$

E_f $\quad = 1/(1 + e^{0}) = 0.500$

$E_f - 0.05$ eV $\quad = 1/(1 + e^{-1.98}) = 0.879$

$E_f - 0.10$ eV $\quad = 1/(1 + e^{-3.96}) = 0.981$

$E_f - 0.15$ eV $\quad = 1/(1 + e^{-5.95}) = 0.997$

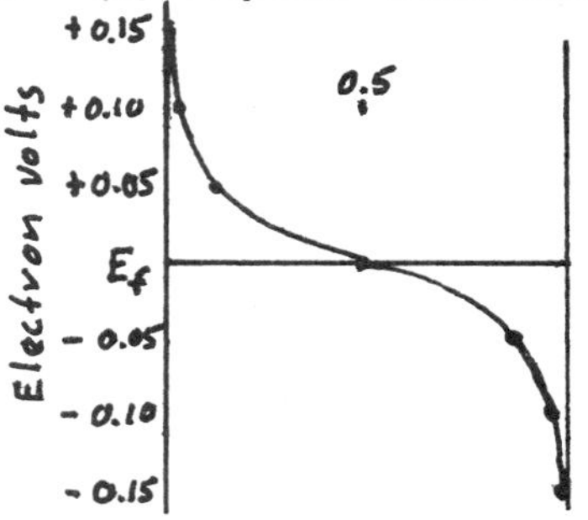

Comment: Note the symmetry. Above E_f, the probability of electrons equals the probability of holes below E_f.

11-P33 Explain to a classmate why there are energies that are forbidden (a) to electrons that are associated with individual atoms; (b) to delocalized electrons that are in metallic structures.

Explanation : (a) Only standing waves will presist for the electrons around the atom Electrons with other wavelengths (other radii about the atom) will be cancelled.

(b) Electrons can only travel in directions that match the wavelength with the atomic spacing. This matching does not exist for all wavelengths. Therefore, some wavelengths are not permitted.

11-P34 Differentiate among *metallic conductors*, *semiconductors*, and *insulators* on the basis of energy bands.

Comparisons : Metallic conductors Valence band is not filled. Electrons are easily raised with thermal energy to overlying vacant energy states.

Insulators There is a large energy gap between the highest filled band and the bottom of the first vacant band. Very few electrons can be energized across this gap.

Semiconductors There is a relatively small energy gap between the highest filled band and the bottom of the first vacant band. A useful number of electrons have the thermal energy to jump this gap and provide conductivity.

11-P41 (a) What fraction of the charge is carried by electron holes in intrinsic gallium arsenide (GaAs)? (b) What fraction by the electrons?

Procedure : The conductivity of an intrinsic semiconductor results from equal numbers of positive and negative carriers. Thus, $\sigma_{in} = \sigma_n + \sigma_p = n_n q\,\mu_n + n_p q\,\mu_p = n_n q\,(\mu_n + \mu_p)$ and $\sigma_n/(\sigma_n + \sigma_p) = \mu_n/(\mu_n + \mu_p)$, since $n_n = n_p$ and q is a constant (0.16 $\times10^{-18}$ A·s)

Solution :(a) $\sigma_p/(\sigma_n+\sigma_p) = \mu_p/(\mu_n+\mu_p)$ = (0.04 m^2/V·s)/(0.88 +0.04 m^2/V·s) = **0.043**

(b) $\sigma_n/(\sigma_n+\sigma_p) = \mu_n/(\mu_n+\mu_p)$ = (0.88 m^2/V·s)/(0.88 +0.04 m^2/V·s) = **0.96**

11-P42 The resistivity of a semiconductor that possesses $10^{21}/m^3$ of negative carriers (plus a negligible number of positive carriers) is 0.016 ohm·m. (a) What is the conductivity? (b) What is the electron mobility? (c) What is the drift velocity when the voltage gradient is 5 mV/mm? (d) What is the drift velocity when the voltage gradient is 0.5 V/m?

Solution : (a) $\sigma = 1/\rho$ = 1/0.016 ohm·m = **62.5 m^2/V·s.**

(b) $\mu = 1/n\,q\,\rho$ = 1/[($10^{21}/m^3$) (0.16 $\times10^{-18}$ A·s)(0.016 ohm·m) = **0.39 m^2/V·s.**

(c_1) $\bar{v} = \mu\mathcal{E}$ = (0.39 m^2/V·s)(0.005 V)/(0.001m) = **1.95 m/s;** (c_2) **0.195 m/s**

11-P43 Based on the data in Table 11-4.2, which is larger, the conductivity from electrons in intrinsic InP, or the conductivity from electron holes in intrinsic InAs?

InP: σ_n = (500 $ohm^{-1}\cdot m^{-1}$)[0.47/(0.47 + 0.015 m^2/V·s)] = **485 m^2/V·s**

InAs: σ_p = (10,000 $ohm^{-1}\cdot m^{-1}$)[0.026/(2.26 + 0.026 m^2/V·s)] = **114 m^2/V·s** .

11-P44 Pure silicon has 32 valence electrons per unit cell (8 atoms with 4 valence electrons each). Its resistivity is 2000 ohm·m. What fraction of the valence electrons are available for conduction?

Procedure : Determine the total number of valence electrons/m^3 from 32/unit cell. Calculate the number of carriers from Eq. (11-1.1). Check your answer with the value in Table 11-4.1.

Solution : N = 32 electrons/(0.543 $\times10^{-9}$ m)^3 = 2 $\times10^{29}/m^3$.

$1/n$ = $(\rho q(\mu_n + \mu_p)$ = (2000 ohm·m)(0.16 $\times10^{-18}$ A·s)(0.19 + 0.0425 m^2/V·s)

n = 1.35 $\times10^{16}/m^3$. n/N = (1.35 $\times10^{16}/m^3$)/(2 $\times10^{29}/m^3$) ≈ **10^{-13}** .

11-P45 The mobility of electrons in silicon is 0.19 m^2/V·s. (a) What voltage is required across a 2-mm chip of Si to produce a drift velocity of 0.7 m/s? (b) What electron concentration must be in the conduction band to produce a conductivity from negative carriers of 20 $ohm^{-1}\cdot m^{-1}$? (c) What would be the total conductivity for pure silicon?

Procedure : Use the basic relationships of Section 11-2.

Solutions : (a) $E = \mathcal{E}d$ = $(\bar{v}/\mu)d$ = (0.7 m/s)(0.002 m)/(0.19 m^2/V·s) = **7.4 mV** .

(b) σ = 20 $ohm^{-1}\cdot m^{-1}$ = n (0.16 $\times10^{-18}$ A·s)(0.19 m^2/V·s) = **6.6 $\times10^{20}$ m^3** .

(c) $\sigma = \sigma_n + \sigma_p$ = (6.6 $\times10^{20}$ m^3)(0.16 $\times10^{-18}$ A·s)(0.19 + 0.0425 m^2/V·s)

= **24.5 ($ohm^{-1}\cdot nm^{-1}$).**

11-P46 The conductivity for silicon is 5 x 10^{-4} $ohm^{-1} \cdot m^{-1}$ at 20°C. Estimate the conductivity at 30°C.

Estimate: From Table 11-4.2, E_g = 1.1 ev, and $\sigma_{20°C}$ = 5 x10^{-4} $ohm^{-1} \cdot m^{-1}$.

ln (5 x10^{-4} $ohm^{-1} \cdot m^{-1}$) = $ln\ \sigma_o$ - (1.1 eV)/[2(86.1 x10^{-6}eV/K)(293 K)]

Solving, $ln\ \sigma_o$ = 14.2; σ_o = 1.5 x10^6 $ohm^{-1} \cdot m^{-1}$

$\sigma_{30°C}$ = (1.5 x10^6 $ohm^{-1} \cdot m^{-1}$) exp[-1.1 eV)/2(86.1 x10^{-6}eV/K)(303 K)]

= **10^{-3} $ohm^{-1} \cdot m^{-1}$.**

11-P47 An intrinsic semiconductor has a conductivity of 390 $ohm^{-1} \cdot m^{-1}$ at 5°C and of 1010 $ohm^{-1} \cdot m^{-1}$ at 25°C. (a) What is the size of the energy gap? (b) What is the conductivity at 15°C?

ln (390 $ohm^{-1} \cdot m^{-1}$) = $ln\ \sigma_o$ - (E_g)/[2(86.1 x10^{-6}eV/K)(278 K)]

ln (1010 $ohm^{-1} \cdot m^{-1}$) = $ln\ \sigma_o$ - (E_g)/[2(86.1 x10^{-6}eV/K)(298 K)]

(a) Solving simultaneously, $ln\ \sigma_o$ = 20.17; and E_g = **0.68 eV** .

(b) $ln\ \sigma_{15°C}$ = 20.17 - 0.68 eV/[2(86.1 x10^{-6}eV/K)(288 K)] = 6.46

$\sigma_{15°C}$ = **640 $ohm^{-1} \cdot m^{-1}$** .

11-P51 Silicon has a density of 2.33 g/cm^3. (a) What is the concentration of silicon atoms per m^3? (b) Phosphorus is added to the silicon to make it an *n*-type semiconductor of 100 $ohm^{-1} \cdot m^{-1}$. What is the concentration of donor electrons per m^3?

Procedure: Part (a) is based on earlier chapters. Part(b) involves negative carriers only.

Solution: (a) (2.33 x10^6 g/m^3)/[(28.06 g)/(0.6 x 10^{24} Si)] = **5 x10^{28} Si/m^3.**

(b) $n = \sigma / q\mu$ = (100 $ohm^{-1} \cdot m^{-1}$)/(0.16 x10^{-18} A·s)(0.19 m^2/V·s) = **3.3 x10^{21}/m^3.**

11-P52 Refer to Example 11-5.1. How many silicon atoms are there for each aluminum atom ?

See Problem 11-P51(a), N_{Si} = 5 x10^{28} Si/m^3 \ Si/Al = **1.7 x10^6**.

From Example 11-P51, N_{Al} = 3 x10^{22} Si/m^3 /

11-P53 Aluminum is a critical impurity when present in silicon during processing for semiconductors. Assume only 10 ppb (0.000,001 a/o). Aluminum remains in the final product. Will the resulting extrinsic semiconductivity be greater than or less than the intrinsic conductivity of silicon at 20°C?

Data: From Table 11-4.2, a_{Si} =0.543 nm; and σ_{in} = **5 x10^{-4} $ohm^{-1} \cdot m^{-1}$** .

Solution: Al/m^3 = [8 Si/(0.543 x10^{-9} m)^3](10^{-8} Al/Si) = 5 x10^{20} /m^3 = n_p .

σ_{ex} = (5 x10^{20} /m^3)(0.16 x10^{-18} A·s)(0.0425 m^2/V·s) = **3.4 $ohm^{-1} \cdot m^{-1}$. (Greater)**

11-P54 Differentiate between *acceptor* and *donor* impurities.

Difference: Acceptor impurities receive electrons from the valence band, thus producing electron holes in the valence band. These holes act as carriers.

Donor impurities donate electrons to the conduction band to become carriers. In neither case are electrons required to jump the full gap to produce conductivity. Furthermore, $n \neq p$.

11-P55 Refer to Problem 11-P53. At what temperature will 1 percent of the conductivity be intrinsic?

Procedure: If 0.99 σ = 3.4 $ohm^{-1} \cdot m^{-1}$, σ_{total} = 3.43; and 0.01 σ =0.0343 $ohm^{-1} \cdot m^{-1}$.

Solution:

$$\ln \sigma_{20°C} = \ln (5 \times 10^{-4}\ ohm^{-1} \cdot m^{-1}) = \ln \sigma_o - (1.1\ eV)/[2(86.1 \times 10^{-6} eV/K)(293\ K)]$$

$$\ln \sigma_T = \ln (0.0343\ ohm^{-1} \cdot m^{-1}) = \ln \sigma_o - (1.1\ eV)/[2(86.1 \times 10^{-6} eV/K) T]$$

Solving: $\ln (5 \times 10^{-4}/0.0343) = -4.23 = [(-1.1\ eV)/[2(86.1 \times 10^{-6} eV/K)][1/293 - 1/T]$

T = 363.5 K = **~90°C.**

11-P56 Detail the mechanism of acceptor saturation in terms of Fig. 11-5.4

Explanation: If the acceptor impurities are limited in number, e.g., 10^{21} Al/m^3 within silicon, essentially all of the acceptor sites can be filled below the temperature of normal usage. No more electron holes can be produced in the valence band by the impuritiies. Thus, the number of carriers and the conductivity remain essentially constant (until a much higher temperature, when a measurable number of electrons jump the full gap).

11-P57 In an extrinsic semiconductor, the Fermi energy moves toward the center of the energy gap as the temperature is increased. Why does this happen?

Explanation: Intrinsic conductivity increases with temperature, because more electrons jump the gap. Gradually they overshadow the extrinsic carriers so that the distribution shifts from that in Fig. 11-5.3 to that in Fig. 11-4.6.

11-P58 Refer to Example 4-5.1. (a) Is the oxide *n*- or *p*-type? (b) How many charge carriers are there per mm^3?

Answer: (a) This has a defect structure because the Fe^{3+} ions lack an electron vis-s-vis the Fe^{2+} ions. The Fe^{3+} ion can accept an electron; therefore, ***p*-type.**

(b) From Example 4-5.1, 8 Fe^{3+} per 13 unit cells, and a = 0.429 nm.

$$n_p = (8\ \text{electron holes})/(13)(0.429 \times 10^{-6}\ mm)^3 = \mathbf{7.8 \times 10^{18}/mm^3}.$$

11-P59 Silicon has one atom of gallium replacing one atom of silicon per 10^6 unit cells. (a) Will the material be *n*- or *p*-type? (b) On the basis of the data in Table 11-4.2, what is the resulting conductivity?

Solution: (a) Gallium is in Group III, therefore ***p*-type.**

(b) $n = 1/(10^6\ u.c.)(0.543 \times 10^{-9}\ m)^3 = 6.25 \times 10^{21}/m^3$.

$$\sigma = nq\mu = (6.25 \times 10^{21}/m^3)(0.16 \times 10^{-18}\ A \cdot s)(0.0425\ m^2/V \cdot s) = \mathbf{42.5\ ohm^{-1} \cdot m^{-1}}.$$

11-P61 The resistance of some silicon for a thermistor is 1031 ohm at 25.1°C. With no change in the measurement procedure, the resistance decreases to 1029 ohm. What is the change in temperature?

11-P61 (con't) *Procedure* : With constant dimensions, $R_2/R_1 = \rho_2/\rho_1 = \sigma_1/\sigma_2$

Solution : R_2/R_1 = 1029 Ω/1031 Ω = $\sigma_1/\sigma_2 = \sigma_{25.1°C}/\sigma_T$.

$$\frac{\sigma_{25.1}}{\sigma_T} = \frac{\sigma_0 \exp[-1.1 \text{ eV})/2(86.1 \times 10^{-6} \text{ eV/K})(298.1 \text{ K})]}{\sigma_0 \exp[-1.1 \text{ eV})/2(86.1 \times 10^{-6} \text{ eV/K})T]} = \frac{R_2}{R_1} = \frac{1029 \ \Omega}{1031 \ \Omega} .$$

Solving simultaneously, T = 298.13 K = 25.13°C; ΔT = **+0.03°C.**

11-P62 To what temperature must you raise GaP to make its resistivity 50 percent of its value at 0°C?

$$\frac{R_2}{R_1} = 0.5 = \frac{\rho_2}{\rho_1} = \frac{\sigma_1}{\sigma_2} = \frac{\sigma_{273}}{\sigma_T} = \frac{\sigma_0 \exp[-2.3 \text{ eV})/2(86.1 \times 10^{-6} \text{ eV/K})(273 \text{ K})]}{\sigma_0 \exp[-2.3 \text{ eV})/2(86.1 \times 10^{-6} \text{ eV/K})T]}$$

Solving simultaneously, T = 277 K = **4°C** .

11-P63 Refer to Example 11-6.2. If the emitter voltage is increased 70 percent from 17 to 29 mV, by what factor is the collector current increased?

Procedure : Use Eq. (11-6.2), and data from Example 11-6.2.

Solution : $\ln I_C$ = -1.17 + (29 mV/6.25 mV) = 3.47

I_C = 32 mA; $(I_C)_{29 \text{ mV}}/(I_C)_{17 \text{mV}}$ = 32 mA/4.7 mA = **6.8**

12-P11 Estimate the maximum BH product of Alnico V on the basis of data in Fig.12-1.3.

Procedure : Pick several H and B values from the 2nd quadrant. Include intermediate points if necessary to locate the maximum.

Solution :

At H = 25,000 A/m,	B = 1.15 V·s/m^2,	BH = 28,750 J/m^3
33,000 A/m,	1.0 ,	BH = 33,000 J/m^3
40,000 A/m,	0.7 ,	BH = 28,000 J/m^3

$BH_{max} \approx$ **33,000 J/m^3** (34,000 J/m^3 is listed in Table 12-1.1.)

12-P21 Why are strain-hardened steels also harder magnetically, and why are annealed ones softer?

Explanation : Strain-hardened steels contain many dislocations, which pin the domain boundaries to retain the magnetization.

Annealed steels have had many of their dislocations removed. Also there may be grain growth, which reduces the amount of grain-boundary area, and thereby eases domain movements.

12-P22 (a) What magnetic characteristics are required for a soft magnet? (b) What structural features optimize these behaviors?

Answers : (a) Low remanent induction, B_r ; low coecive force, $-H_c$; low energy product, BH_{max}.

(b) Few dislocations (annealed); minimum grain boundaries (grain growth); no phase boundaries (single-phase materials).

12-P23 Estimate the amount of energy required to demagnetize an Alnico V magnet that has a volume of 2.1 cm^3.

Solution : From Table 12-1.1, BH_{max}= 34,000 J/m^3.

Energy = (34,000 J/m^3)(2.1 x10^{-6} m^3) = **0.07 J.**

Comment: This is an index. We would need to integrate the area under the curve in the 2nd quadrant for a full measure of the energy.

12-P31 The ceramic magnetic material nickel ferrite has eight [$NiFe_2O_4$] formula weights per unit cell, which is cubic with a = 0.834 nm. Assume that all the unit cells have the same magnetic orientation. What is the saturation magnetization?

(The Ni^{2+} ion lost both 4s electrons, but retained all 3d electrons, when it ionized.)

Procedure : Cf. Fe_3O_4, Section 12-3.

Solution : β/u.c. = $8(5+)_{Fe^{3+}}$ + $8(5[)_{Fe^{3+}}$ + $8(2+)_{Ni^{2+}}$ = 16 β/u.c. +

M_s = (16 β)(9.27 x10^{-24} A·m^2/β)/(0.834 x10^{-9}m)^3 = **0.25 x10^6 A/m.**

12-P32 Electron hopping occurs from Fe^{2+} to Fe^{3+} ions in iron oxide (Fig 4-5.3 and Problem 115.11). Consider Fe_3O_4, or $[Fe^{2+}Fe_2{}^{3+}O_4]_8$, in which one electron per unit cell has hopped from an Fe^{2+} ion to an Fe^{3+} ion that is in a 4-f site. (a) What is the resulting magnetization per unit cell? (b) What would be the effect if it had hopped from an Fe^{2+} ion to an Fe^{3+} ion that was in a 6-f site?

Procedure : Refer to the table in the text of Section12-3. Make the necessary adjustments.

Solution : (a)

	6-*f*	4-*f*	$\Uparrow = 5\,\beta$; $\uparrow = 4\,\beta$.
	Fe^{3+} $\Uparrow\Uparrow\Uparrow\Uparrow\Uparrow\Uparrow\Uparrow\Uparrow$	Fe^{3+} $\Downarrow\Downarrow\Downarrow\Downarrow\Downarrow\Downarrow\Downarrow\downarrow$ Fe^{2+}	
	Fe^{3+} $\Uparrow\uparrow\uparrow\uparrow\uparrow\uparrow\uparrow\uparrow$ Fe^{2+}		
	$9(5\,\beta) + 7(4\,\beta) = 73\,\beta \Uparrow$	$7(5\,\beta) + 1(4\,\beta) = 39\,\beta \Downarrow$	Net = **34 β $\Uparrow$/u.c.**
(b)	$8(5\,\beta) + 8(4\,\beta) = 72\,\beta \Uparrow$	$8(5\,\beta) + 0(4\,\beta) = 40\,\beta \Downarrow$	Net = **32 β $\Uparrow$/u.c.**

12-P33 Refer to the calculations for the magnetization of Fe_3O_4 in Section 12-3. Al^{3+} ions can substitute for Fe^{3+} ions in the 6-f sites of the magnetite to give $[Fe(Fe,Al)_2O_4]_8$ as a solid solution. If one-half of these Fe^{3+} ions are replaced by Al^{3+} ions, what is the resulting magnetization? (Assume that the Fe^{3+} ions in the 4-f sites are unaffected.)

Solution :

	Fe^{2+} ($4\,\beta$)	Fe^{3+} ($5\,\beta$)	Al^{3+} ($0\,\beta$)	$\Sigma\beta$	
6-*f* $\uparrow$	8			+32	\
6-*f* $\Uparrow$		4	4	+20	Net = +12 β
4-*f* $\Downarrow$		8		-40	/

Assume that the unit cell does not change in size.

$$M = \Sigma p_m/V = (12\,\beta)(9.27 \times 10^{-24}\ \mathrm{A{\cdot}m^2/\beta})/(0.84 \times 10^{-9}\mathrm{m})^3 = \mathbf{0.2 \times 10^6\ A/m.}$$

12-P41 Metallic cobalt has a saturation magnetization of 1.43×10^6 $\mathrm{A{\cdot}m^2/m^3}$. Calculate the magnetic moment per atom (a) in $\mathrm{A{\cdot}m^2}$, and (b) in Bohr magnetons.

Procedure : Determine the number of atoms/m^3 as we did in earlier chapters. From this the values per atom. Data are from the appendix.

Solution : $(8.9 \times 10^6\ \mathrm{g/m^3})(0.6 \times 10^{24}\ \mathrm{Co}/58.93\ \mathrm{g}) = 9.1 \times 10^{28}\ \mathrm{Co/m^3}$

(a) $(1.43 \times 10^6\ \mathrm{A{\cdot}m^2/m^3})/(9.1 \times 10^{28}\ \mathrm{Co/m^3}) = \mathbf{1.57 \times 10^{-23}\ A{\cdot}m^2/Co}$

(b) $(1.57 \times 10^{-23}\ \mathrm{A{\cdot}m^2/Co})/(9.27 \times 10^{-24}\ \mathrm{A{\cdot}m^2/\beta}) = \mathbf{1.7\ \beta/Co}$

12-P42 (a) Distinguish between *ferrimagnetism* and *antiferromagnetism*. (b) Distinguish between *paramagnetism* and *ferromagnetism*.

Contrasts : Ferrimagnetism. Magnetic moments of atoms are aligned in opposing directions, but unbalanced. Thus, a net magnetization is observed externally to the material.

12-P42 (con't) Antiferromagnetism. Magnetic moments of atoms are aligned in opposing directions, but balanced. Net magnetization is zero.

Paramagnetism. Individual atoms possess a magnetic moment; however there is no coupling into a common direction until an external field is applied.

Ferromagnetism. Metallic materials with spontaneous alignment of electrons within the valence band.

12-P43 Refer to Example 12-3.1. Assume that 40 percent of the domains have their N---S polarity reversed 180° after the magnetic field H_c drops to zero. What is the remanent induction, B_r?

Solution : Full magnetization is 600,000 A/m. 240,000 have reversed; 360,000 have not.

Therefore, 360,000 A/m + (- 240,000 A/m) = **120,000 A/m** .

Comment : The 570,000 A/m value cited in Example 12-3.1 is a result of a nonalignment of 2-3 percent of the domains.

13-P21 Two capacitor plates (20 mm x 30 mm each) are parallel and 2.2 mm apart, with nothing between them. What voltage is required to develop a charge of 0.24 x 10^{-10} C on the electrodes?

Solution : Rearranging Eqs. (13-2.3 and 13-2.4), $E = (Q/A)/(\epsilon_0/d)$.

$$E = [(0.24 \times 10^{-10}\text{ C})/(0.03\text{ m} \times 0.02\text{ m})]/[8.85 \times 10^{-12}\text{ C/V·m})/(0.0022\text{ m})] = \mathbf{10\ V.}$$

13-P22 What is the electron density on the electrodes in Problem 13-P21?

Solution : $\text{El./m}^2 = [(0.24 \times 10^{-10}\text{C})/(0.03\text{ m} \times 0.02\text{ m})]/(0.16 \times 10^{-18}\text{ C/el})$
$= \mathbf{250 \times 10^9\ el/m^2}$

13-P23 A 2.2-mm sheet of polystyrene is inserted in the space between the plates of Problem 13-P21. With 10 V, the charge is 0.24 x 10^{-10}C without the polymer sheet, and 0.6 x 10^{-10}C with the sheet. What is the relative dielectric constant of the plastic?

Solution : The dielectric constant is the ratio of the charge density with a dielectric present to the charge density with no dielectric present.

$$\kappa = \mathcal{D}_m/\mathcal{D}_0 = (0.6 \times 10^{-10}\text{ C})/(0.24 \times 10^{-10}\text{ C}) = \mathbf{2.5}\ .$$

13-P31 Dielectric constants of polymers are greatest at intermediate temperatures. Why is this so? How does the dielectric constant differ between ac and dc.

Explanation : There can be no molecular polarization until the temperature is raised above T_g, when the polar groups can respond to the electric field. At still higher temperatures, thermal agitation destroys the dipole orientation, reducing the polarization and the dielectric constant. Higher with **dc**. There is ample time for polar groups to reorient. Higher ac frequencies reduce the amount of polarization and therefore the relative dielectric constant.

13-P32 A plate capacitor must have a capacitance of 0.25 μf. What should its area be if the 0.0005-in. (0.0131-mm) mylar film that is used as a spacer has a dielectric constant of 3.0? (See Example 13-3.1.)

Solution : Since 1μf = 10^{-6} C/V, and using Eq. (13-3.1),

$$A = (0.25 \times 10^{-6}\text{C/V})(13.1 \times 10^{-6}\text{m})/(3.0)(8.85 \times 10^{-12}\text{ C/V·m})$$
$$= 0.12\text{ m}^2, \quad \text{or } \mathbf{1200\ cm^2}.$$

13-P41 A piezoelectric crystal has an elastic modulus of 19,000,000 psi (130 GPa). What stress will reduce its polarization from 560 to 557 C/m^2?

Procedure : The change in polarization, $\Delta\mathcal{P}/\mathcal{P} = (\Delta Qd/V)/(Qd/V)$, depends entirely on Δd, since the other values are constant. This Δd must be produced by the strain, e .

Therefore, $\Delta\mathcal{P}/\mathcal{P} = s/E$.

Solution : $s = [(557 - 560\text{ C/m}^2)/(560\text{ C/m}^2)](19{,}000{,}000\text{ psi}) =$ **-10^5 psi.** (Comp.)

Or $s = [(557 - 560\text{ C/m}^2)/(560\text{ C/m}^2)](130{,}000\text{ MPa}) =$ **−700 MPa.** (")

13-P42 Refer to Example 13-4.2. What is the distance between the centers of positive and negative charges in each unit cell?

Procedure : Use the center plane of the Ba^{2+} positions for reference.

Solution : Center of + : $[(+2)(0\ ) + (+4)(0.006\ nm)]/(6+) = +0.0040\ nm$

Center of - : $[(-4)(0.006\ ) + (-2)(0.008\ nm)]/(6-) = \underline{-0.0067\ nm}$

$d = \Delta = $ **0.0107 nm**

13-P43 Lead zirconate is cubic in one of its polymorphs. The unit cell can be chosen so that each corner has a Zr^{4+} ion; the center of an edge, an O^{2-} ion; and the center of the unit cell, a Pb^{2+} ion.
(a) What is the chemical formula? (b) Relocate the unit cell so that the Zr^{4+} ion is at the center. Speculate on ferroelectric possibilities.

Solution : (a) 1 Pb^{2+}, $^8/_8$ Zr^{4+}, $^{12}/_4$ O^{2-} = **$PbZrO_3$.** (b)

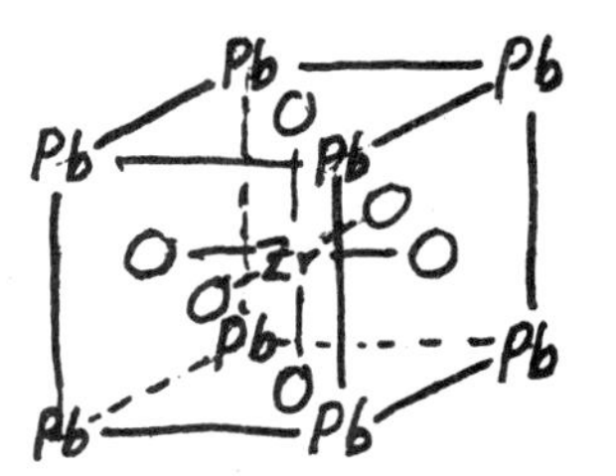

(c) It has the same structure as $BaTiO_3$ and $CaTiO_3$ (Fig. 3-2.9). The cubic form can not be ferroelectric because the centers of the positive and negative charges are coincident. To be ferroelectric, it would have to become noncubic, as does $BaTiO_3$ below 120°C (Fig. 13-4.5). This change occurs in $PbZrO_3$ at 233°C.

13-P44 Explain basis for the ferroelectric Curie temperature in terms of crystal structure.

Explanation : **Below** the ferroelectric Curie temperature (120°C for $BaTiO_3$): The Ti^{4+} ions and the O^{2-} ions are not symmetrically placed with respect to the center of the unit cell (based on Ba^{2+}). Figure 13-4.5 shows two such mirror-image displacements. (In three dimensions, there are six.)

Above the ferroelectric Curie temperature: Available thermal energy lets these ions resonate between these locations, so that on the average, their centers of gravity are at the center of the unit cell (Fig. 3-2.9).

13-P45 Microwave heating involves molecular polarization. Explain this statement.

Explanation : The H_2O molecule is polar, since the hydrogen atoms are simply unscreened protons at the end of a covalent bond (Fig. 2-2.5b). The molecule reorients with the alternating field at microwave frequencies. Some of the energy input is dissipated as heat, particularly if the frequency is near the limit of the reversal times for the water molecules. Other polar groups that are present respond similarly.

13-P51 **The critical angle for internal reflection at a glass-air surface is 41°. (a) What is the index of the glass? (Assume that $n_{air} = n_{vac}$.) (b) What is the velocity of light in the glass?**

Solution : (a) $n_{glass} / n_{air} = \sin \phi_{air} / \sin \phi_{glass} = n_{glass} / 1.0 = \sin 90° / \sin 41°$.

$n_{glass} =$ **1.524** .

(b) $v_{glass} = v_0 / n_{glass} = (3 \times 10^8 \text{ m/s})/1.524 =$ **1.97×10^8 m/s** .

13-P52 **Light retains 60 percent of its intensity after traveling 600 m through a glass fiber. What is its attenuation in db/km?**

Solution : Based on Eq. (13-5.4), $0.1 \log_{10}(0.6/1) = 0.1(-0.22)$ db/600m

-0.022 db/600 m $= -37 \times 10^{-6}$ db/m $=$ **-3.7 db/km** .

13-P53 **Two spectral lines have wavelengths of 589.3 and 656.3 nm. Their indices of refraction in an optical glass are 1.516 and 1.501, repectively. How much will their refracted angles differ, if the incident ray of each is 19°30' from the normal as they enter the glass from the air?**

Solution : $n_{air} / n_{glass} = \sin \phi_{glass} / \sin \phi_{air}$

Yellow: $\sin \phi_{589.3} = (1/1.516)(\sin 19.5°) = 0.2202$; $\phi = 12.72°$, or **12°43'** .

Red: $\sin \phi_{656.3} = (1/1.501)(\sin 19.5°) = 0.2224$; $\phi = 12.85°$, or **12°51'** .

13-P61 **A phosphorescent material is exposed to ultraviolet light. The intensity of the reemitted light decreases by 20 percent in the first 37 min after the ultraviolet light is removed. (a) How long will it be after the ultraviolet light has been removed before the light has only 20 percent of its original intensity (a decrease of 80 percent)? (b) How long before it has only 1 percent?**

Solution : $I/I_0 = 0.8 = e^{-37/\lambda}$ $\lambda = 166$ min

(a) $= 0.2 = e^{-t/166}$ $t =$ **267 min**

(b) $= 0.01 = e^{-t/166}$ $t =$ **764 min** .

13-P62 **A phosphorescent material must have an intensity of 50 (arbitrary units) after 24 hr and of 20 after 48 hr. Based on these figures, what initial intensity is required? (Solve *without* using a calculator.)**

Solution (without a calculator): For 24 → 48 hr, ratio = 50/20 = 2.5 .

For 0 → 24 hr (same time span), ratio = 2.5 = I_0 /50. $I_0 =$ **125** .

14-P21 How many coulombs (A·s) are required to plate each 1 g of nickel from an Ni^{2+} electrolyte?

Procedure : Basis: 1 g. Determine the number of atoms. Each involves two electrons.

Solution : $(0.6 \times 10^{24}$ Ni/58.7 g)(2 el/Ni)(0.16 $\times 10^{-18}$ A·s/el) = **3300 C/g** .

14-P22 A metallic reflector that has 1.27 m^2 of surface is being chrome plated. The current is 100 A. (a) How many g of Cr^{2+} must be added to the electrolyte per hour of plating? (b) What thickness will accumulate per hour?

Procedure : Determine the number of electrons required; then the number of Cr^{2+}.

Solution : (a)
$$\frac{(100\ \text{A})(3600\ \text{s/hr})(1\,Cr^{2+}/2\ \text{el})}{(0.16 \times 10^{-18}\ \text{A·s/el})(0.6 \times 10^{24}\,\text{Cr}/52\ \text{g Cr})} = \mathbf{97.2\ g/hr}\ .$$

(b) $[(97.2\ \text{g/hr})/(7.2\ \text{g/m}^3)]/(1.27\ \text{m}^2) = 10^{-5}$ m = **0.01 mm** .

14-P23 The electrodes of a standard galvanic cell are nickel and magnesium. What potential difference will be established?

Procedure : Compare them with the standard $H_2 \longrightarrow 2\,H^+ + 2\,e^-$ potential.

Solution :

Mg (anode) - H_2 (cathode) -2.36 V

Ni (anode) - H_2 (cathode) -0.25 V

Mg (anode) - Ni (cathode) **-2.1 V** (with Mg anodic) .

14-P24 Why do we use the hydrogen electrode as the reference electrode rather than one of the metals (e.g., lead)?

Rationale : Hydrogen and H_2O are ubiquitous. Thus, aqueous solutions are widely encountered in corrosion situations. Also, aqueous solutions are readily standardized in the laboratory.

14-P25 Distinguish between *anode* and *cathode* on the basis of electron movements. How does the cathode ray tube fit into the definitions of Section 14-2?

Answers : Anodes provide electrons to the external circuit.

Cathodes receive electrons from the external circuit.

In a CRT, the electrons move internally form the hot filament to the anodized viewing surface. From there, they leave the CRT. (Current moves in the opposite direction.)

14-P26 What copper concentration (g/l) is required in an electrolyte for it to have an electrode potential of 0.32 V (with respect to hydrogen)?

Procedure : This figure is lower than the +0.34 V found in Table 14-2.1, indicating that the concentration is less than the standard 1-molar solution. Use the Nernst equation.

Solution : 0.32 V = +0.34 V + (0.0257 V/2)$\ln C$ $\qquad C$ = 0.21 mol/l;

(0.21 mol/l)(63.54 g/mol) = **13.4 g/l** .

14-P27 **In Fig. 14-2.2, copper is replaced by gold and zinc by tin. Based on the half-cell reactions of Table 14-2.1, will the weight of tin corroded per hr be greater or less than the weight of the gold that is plated?**

Solution :

$$m_{Au}/m_{Zn} = \frac{It(197.0 \text{ amu/Au})(1\text{Au}/3\text{el})/(0.16 \times 10^{-18} \text{ C/el})}{It(118.7 \text{ amu/Sn})(1\text{Sn}/2\text{el})/(0.16 \times 10\text{-}18 \text{ C/el})} = 1.1$$

or +10%

14-P28 **What is the source of the electrons in an ordinary, old-fashioned dry cell? (The electrolyte is a gelatinous paste containing NH_4Cl. The cathode reaction changes Mn^{4+} ions in MnO_2 to Mn^{2+}.)**

Answer : Of course, there must be a second electrode. A quick search discovers that the "cans" of those dry cells are made of zinc. Therefore, $Zn \longrightarrow Zn^{2+} + 2\ e^-$ supplies the electrons.

14-P29 **Consider Eqs. (14-2.2b, 14-2.4, 14-2.5, 14-2.6), and the corrosion of iron. Determine which, if any of, these, provides the prevalent *cathode* reaction if (a) iron is in the water along a seashore, (b) iron is in a copper sulfate solution, (c) iron is in the can containing a carbonated beverage, (d) iron is in the blade of a rusty spade, (e) iron is a car fender that has been scratched.**

Comment: The cathode reaction must involve reactants that are prevalent.

Answers :

(a) Seashore:	$O_2 + 2\ H_2O + 4\ e^- \longrightarrow$	$4(OH)^-$	Eq.(14-2.5)
(b) $CuSO_4$:	$Cu^{2+} + 2\ e^- \longrightarrow$	Cu	Eq.(14-2.2b)
(c) H_2CO_3 (closed):	$2\ H^+ + 2\ e^- \longrightarrow$	$H_2\uparrow$	Eq.(14-2.4)
H_2CO_3 (open):	$O_2 + 2\ H^+ + 2\ e^- \longrightarrow$	$2\ H_2O$	Eq.(14-2.6)
(d) Spade:	$O_2 + 2\ H_2O + 4\ e^- \longrightarrow$	$4(OH)^-$	Eq.(14-2.5)
(e) Fender:	$O_2 + 2\ H_2O + 4\ e^- \longrightarrow$	$4(OH)^-$	Eq.(14-2.5)

14-P31 **With a current density of 0.1 A/m^2, how long will it take to corrode an average of 0.1 mm from the surface of aluminum?**

Procedure : First, determine the electron flux; then the number required per m^2.

Solution : $[(10^{-4}\text{m})(2.7 \times 10^6 \text{g/m}^3)]/[(27 \text{ g}/0.6 \times 10^{24} \text{ Al})(1 \text{ Al}/3 \text{ el})] = 1.8 \times 10^{25} \text{ el/m}^2$

$[(1.8 \times 10^{25} \text{ el/m}^2)(0.16 \times 10^{-18} \text{ A·s/el})/(0.1 \text{ A/m}^2) = 2.9 \times 10^7\text{s} = \sim 11$ mo.

14-P32 **Pit corrosion punctured the hull of an aluminum boat after 12 months in seawater. The aluminum sheet was 1.1 mm thick. The average diameter of the hole was 0.2 mm. (a) How many atoms were removed from the pit per sec? (b) What was the corrosion current density?**

Procedure : Set up your equation by following the dimensional units.

Solution :(a)

$$\frac{[1.1 \text{ mm}(\pi/4)(0.2 \text{ mm})^2(2.7 \times 10^{-3} \text{ g/mm}^3)]}{365 \text{ da}(24 \text{ hr/da})(3600 \text{ s/hr})[27 \text{ g}/(0.6 \times 10^{24} \text{ Al})]} = 6.6 \times 10^{10} \text{ Al/s}.$$

(b) $[(6.6 \times 10^{10} \text{ Al/s})(3 \text{ el/Al})(0.16 \times 10^{-18}\text{C/el})]/[(\pi/4)(0.2 \text{ mm})^2] = 10^{-6} \text{ A/mm}^2$

14-P33 A discarded tin-coated can is left on the shoreline of a local freshwater lake. As rusting proceeds, Eq. (14-2.1$_{Fe}$) is the anodic reaction.

(a) What is the probable cathodic reaction?

(b) Why is the answer to part (a) *not* ($Sn^{2+} + 2e^- \longrightarrow Sn$) ?

Answer : (a) Eq.(14-2.5) $O_2 + 2H_2O + 4e^- \longrightarrow 4(OH)^-$

(b)No tin in the water. ↑air ↑water ↑from anode via cathode

14-P34 Explain why metal in an oxygen-deficient electrolyte becomes anodic to metal in an oxygen-enriched electrolyte.

Explanation : According to the mass law, a deficiency of oxygen in the reaction, $O_2 + 2H_2O + 4e^- \longrightarrow 4(OH)^-$, leads to an increased demand for electrons. (The activity of water remains at unity. $K = a^4_{OH^-}/p_{O_2}v_e^4$).

14-P35 How do barnacles accelerate corrosion on a ship hull?

Explanation : An anode is established in the oxygen deficient area under the barnacle.

14-P36 An enterprising mechanic suggests the use of a magnesium drain plug for the crankcase of a car as a means of avoiding engine corrosion, specifically, of the bearings and cylinder walls. Discuss this proposal.

Comments : +Magnesium is anodic to the steel of the engine parts.
+ Plug is replaceable
- The areas to be protected are not near the drain plug at the bottom of the oil pan.
- If the plug corrodes as a sacrifical anode, leakage follows.

14-P37 Undercoatings of various types are applied to new cars. Under what conditions are they helpful? Under what conditions are they detrimental?

Comments: + Helpful if the coating keeps brines away from the metal surfaces, particularly cracks and crevices
- Harmful if the coating cracks or peels to admit water or brine into areas with low oxygen access. An oxygen concentration cell would result and accelerate corrosion.

14-P41 Sometimes people assume that, if failure does not occur in a fatigue test in 10^8 cycles, the stress is below the endurance limit. The test machine is connected directly to a 1740-rpm motor. How long will it take to log that number of cycles?

Solution : 10^8 cycles/[(1740/min)(60 min/hr)(24 hr/da)] = **40 days** (& 40 nights).

14-P42 Examine the crankshaft of a car. Point out specification and design features that alter the resistance to fatigue.

Observations : Fillets at the edge of the bearing surfaces. Smoother finish on the bearing surfaces. And, of course, the engineering calculations for operating stresses.

14-P51 (a) Why do aluminum kitchen utensils corrode less readily than do iron ones? (b) Why does chromium make steel "stainless"?

Explanation : (a) A protective Al_2O_3 surface film is established (Fig. 14-3.1).

(b) Chromium produces a *passive* surface on stainless steel in the presence of available oxygen. This passive surface is either a chromium oxide layer, or an absorbed oxygen film that acts as a barrier between the metal and the electrolyte. In an oxygen-deficient electrolyte (e.g., hydrochloric acid, HCl), the metal is *activated* and corrosion proceeds rapidly.

14-P52 The following data were obtained in a creep-rupture test of Inconel "X" at 815°C (1500°F): 1 percent strain after 10 hr, 2 percent strain after 200 hrs, 4 percent strain after 2000 hr, 6 percent strain after 4000 hr, "neck down" started at 5000 hr, and rupture occurred at 5500 hr. (a) Sketch the *e*-*t* curve. (b) What is the second-stage creep rate?

(a) *Sketch* :

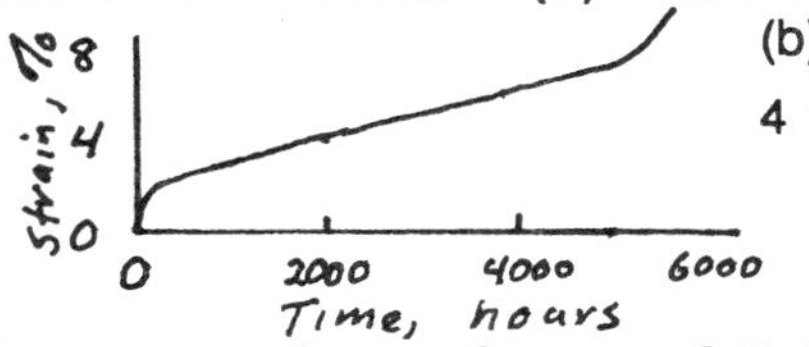

(b) Linear between ~200 and 4000 hrs.

4 % /3800 hr ≈ **0.001 %/hr**

14-P61 A stress relaxes from a 0.7 MPa to 0.5 MPa in 123 days. (a) What is the relaxation time? (b) How long would it take to relax to 0.3 MPa?

Solution : (a) $\ln(s/s_0) = -t/\tau = \ln(0.5/0.7) = -123/\tau$; τ = **366 da .**

(b) $\ln(0.3/0.5) = -t/366$ da $\qquad t$ = **187 additional days**

or $\ln(0.3/0.7) = -t/366$ da $\qquad t$ = **310 total days**

14-P62 An initial stress of 10.4 MPa (1500 psi) is required to strain a piece of rubber 50 percent. After the strain has been maintained constant for 40 days, the stress required is only 5.2 MPa (750 psi). What would be the stress required to maintain the strain after 80 days? Solve this problem without using a calculator.

Solution : In 40 days, $S_{40}/S_0 = 0.50$. Therefore 40 more days, 50% of 50%.

S_{80} = 0.50(750 psi) = **375 psi**; or, S_{80} = 0.25(10.4 MPA) = **2.6 MPa** .

Comment: In this problem, we do not need to calculate that the relaxation time is 58 days.

14-P63 Raw polyisoprene (i.e., nonvulcanized natural rubber) gains 2.3 w/o by becoming cross-linked by oxygen in the air. What fraction of the possible cross-linkes are established?
(In this case, assume that the O_2 cleaves and that each cross-link involves a single oxygen.)

Procedure : As indicated in the legend of Fig. 4-3.10(b), each cross-link may involve a single sulfur (or oxygen) atom rather than two. Therefore with two cross-links per pair of isoprene mers, the saturated ratio, oxygen/isoprene is $2(Ox)/2(C_5H_8)$.

Solution : $2(Ox)/2(C_5H_8)$ = 32 amu/136 amu , or 23.5 amu oxygens/100 amu isoprene.

(2.3 amu Ox)/(23.5 amu Ox) = **0.10**, or 10 percent .

<u>14-P81</u> The average energy of a C--Cl bond is 340 kJ/mol according to Table 2-2.2. Will visible light [400 nm (violet) to 700 nm (red)] have enough energy to break one of these bonds?

Solution : Calculate for blue light, which has the higher energy. Since $E = hc/\lambda$,

$$E = (0.662 \times 10^{-33}\ \mathrm{J{\cdot}s})(3 \times 10^{8}\ \mathrm{m/s})/(400 \times 10^{-9}\ \mathrm{m}) = 5 \times 10^{-19}\ \mathrm{J/bond}$$

$(5 \times 10^{-19}$ J/bond$)(0.6 \times 10^{24}$ bonds/mol$)$ = 300 kJ/mol by blue light. <u>Not enough</u>.

Alternative : $\lambda < [(0.662 \times 10^{-33}\ \mathrm{J{\cdot}s})(3 \times 10^{8}\ \mathrm{m/s})]/[340{,}000\ \mathrm{J}/(0.6 \times 10^{24})$

$<3.5 \times 10^{-7}$ m, or 350 nm, which is ultraviolet . <u>Not enough energy</u> in the visible range.

Comment: However, some bonds will be broken, since there will be some thermal energy present when a photon "hits" the bond.

14-P82 (a) What frequency and wavelength must a photon have to supply the energy necessary to break the average C--H bond in poyethylene? (b) Why can some bonds be broken with slightly longer electromagnetic waves?

Solution : From Table 2-2.2, $E_{C-H} = 435{,}000\ \mathrm{J}/(0.6 \times 10^{24}$ bonds/mol$) = h\upsilon = hc/\lambda$.

(a) $\upsilon = [435{,}000\ \mathrm{J}/(0.6 \times 10^{24}]/[0.662 \times 10^{-33}\ \mathrm{J{\cdot}s}) = \mathbf{1.09 \times 10^{15}/s}$

$\lambda = c/\upsilon = (3 \times 10^{8}\ \mathrm{m/s})/(1.09 \times 10^{15}/\mathrm{s}) = 275 \times 10^{-9}\ \mathrm{m} =$ **275 nm** (ultraviolet)

(b) Thermal energy is also present.

UNITS

for

SELF STUDY

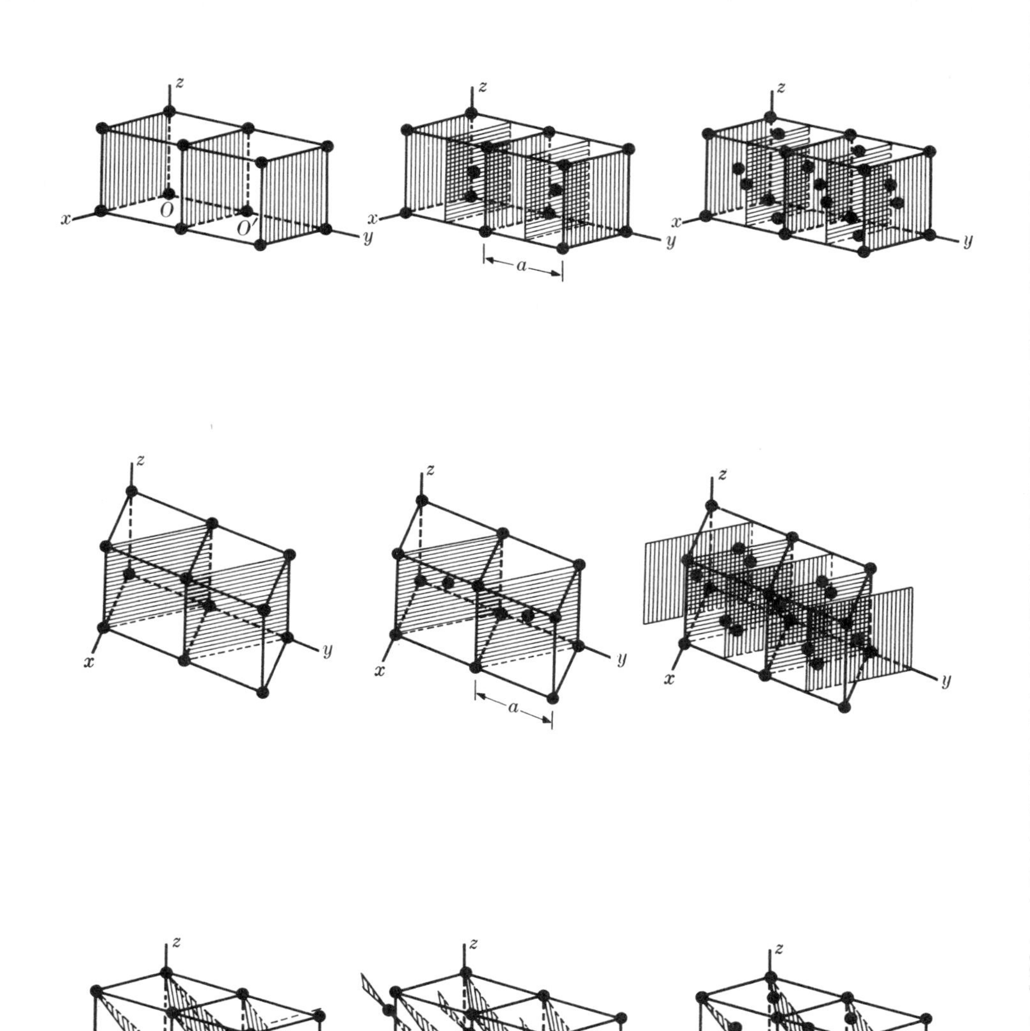

z
x
O
O′
y
a

I

CRYSTAL DIRECTIONS

and

PLANES

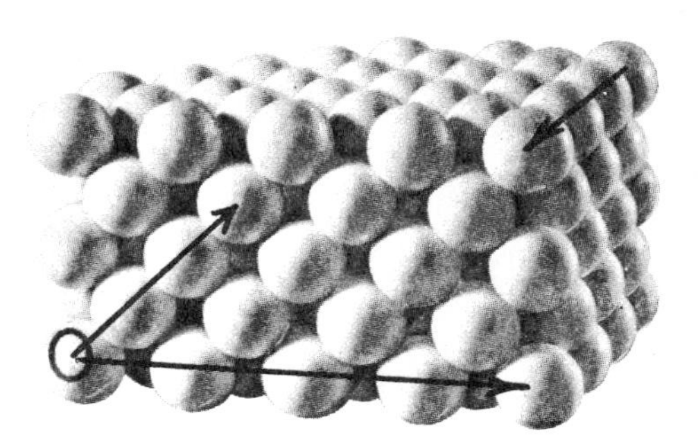

1. **Crystal geometry** (opposite). Many properties of crystals depend on the arrangement of the atoms within the crystals. We can see in the opposite figures that the arrangements differ along different *directions* and on different *planes*. The purpose of this study set is to show the standard ways that crystals are indexed, so we may have a basis for technical communication. (Models by G. R. Fitterer in B. Rogers, *The Nature of Metals*, American Society for Metals.)

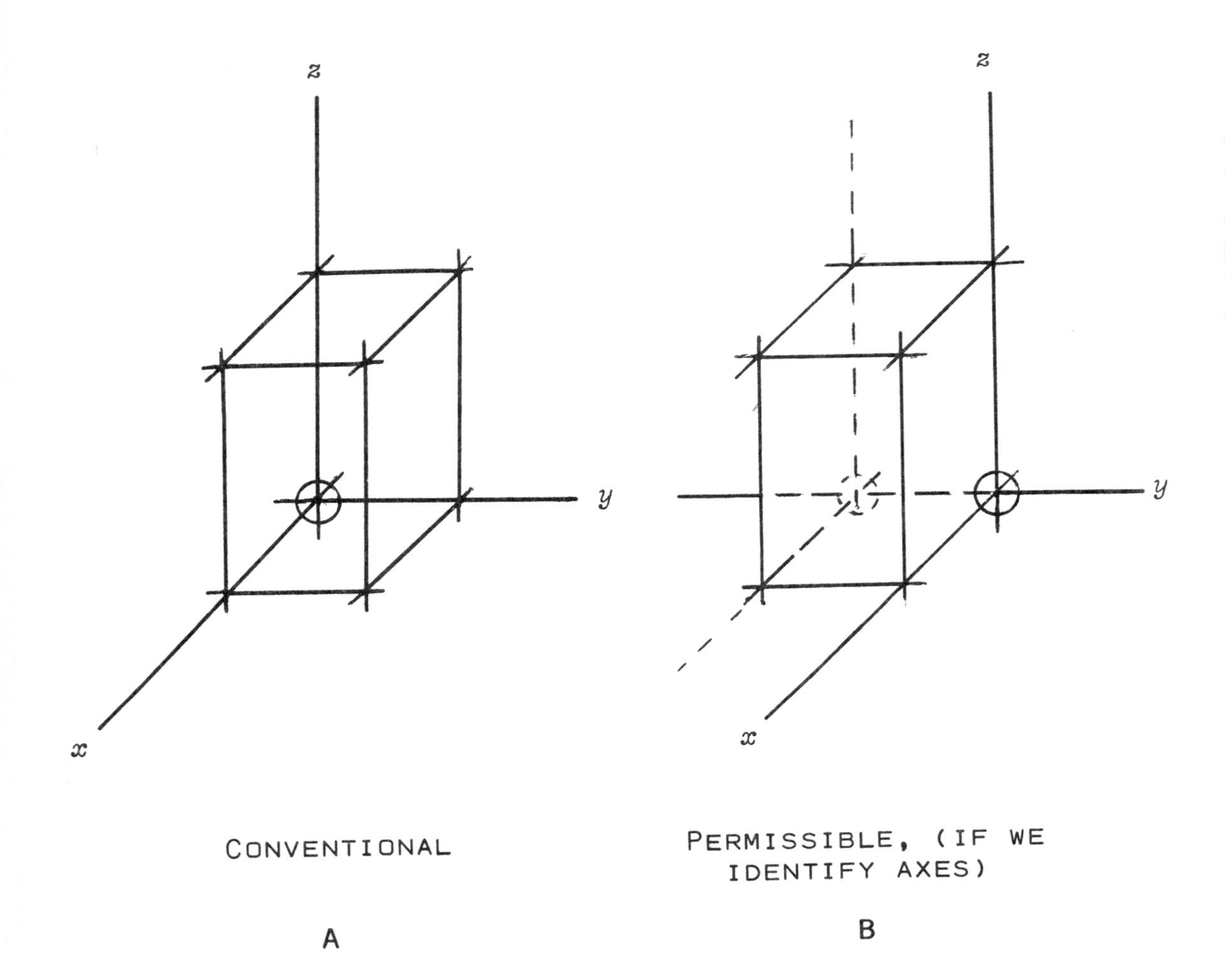

2. **Crystal axes.** For crystal indices, the convention has been established to place the origin at the lower, left, rear corner with the *x*-, *y*-, and *z*-axes oriented as indicated. We are not bound by this convention; however, if we deviate from it, we should specifically indicate the modification to others who are examining our figures. It will save misunderstanding (B). If there is no indication otherwise, we assume that we are using the standard convention (A).

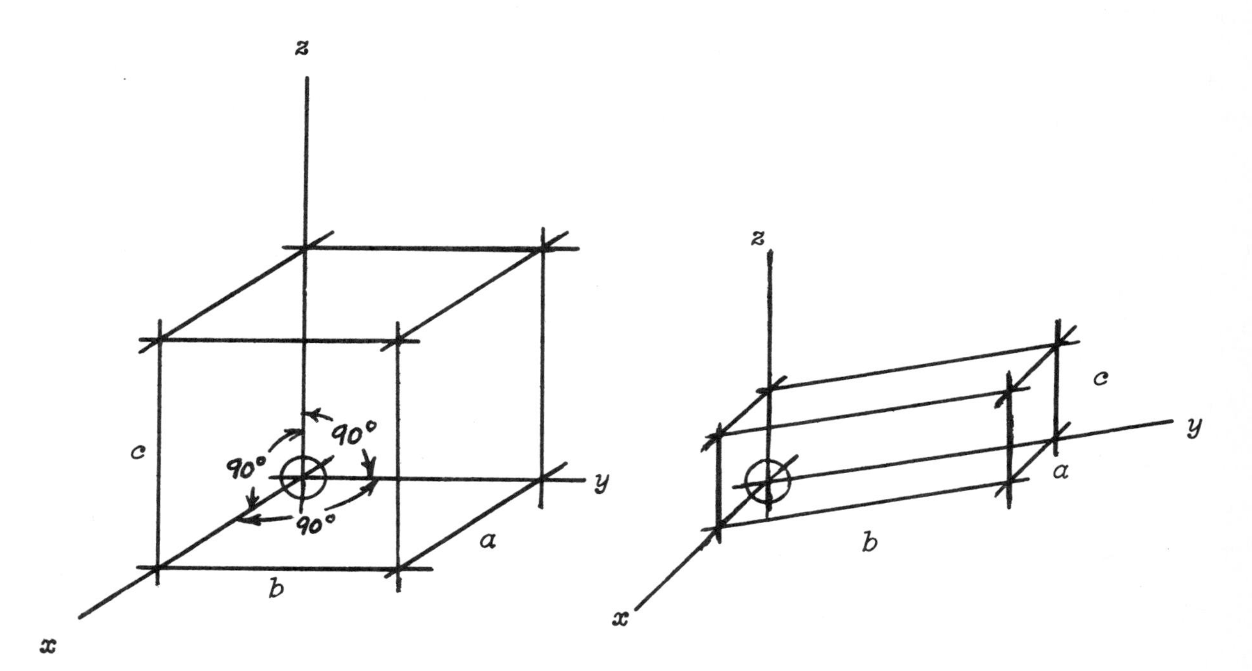

CUBIC	NONCUBIC
$a = b = c$	$a \neq b \neq c$
A	B

3. **Crystal axes** (con't). In *cubic* crystals, the axes are at right angles, and the repeating distances are all equal along the three axes (A). In *non-cubic* crystals, either the axes or the repeating distances, or both, are unlike. (B). Fortunately for indexing crystals, we do not need to know the extent of the change from the cubic structure, because we are indexing on the basis of the unit cell, and not on the basis of meter or nanometer dimensions.

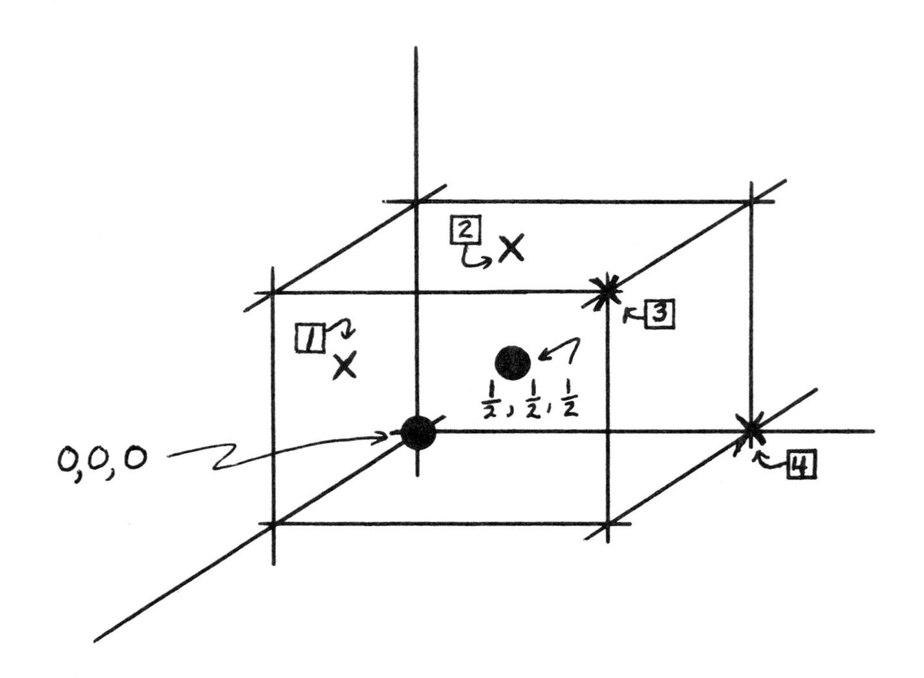

4. **Unit cell locations.** First let us check crystal locations. The *origin* is labeled 0,0,0; the center is labeled by its unit cell coefficients of $^1/_2, ^1/_2, ^1/_2$. Will you identify the location marked [1]; marked [2]; and marked [3]? Check yourself with the answers given on the next figure.

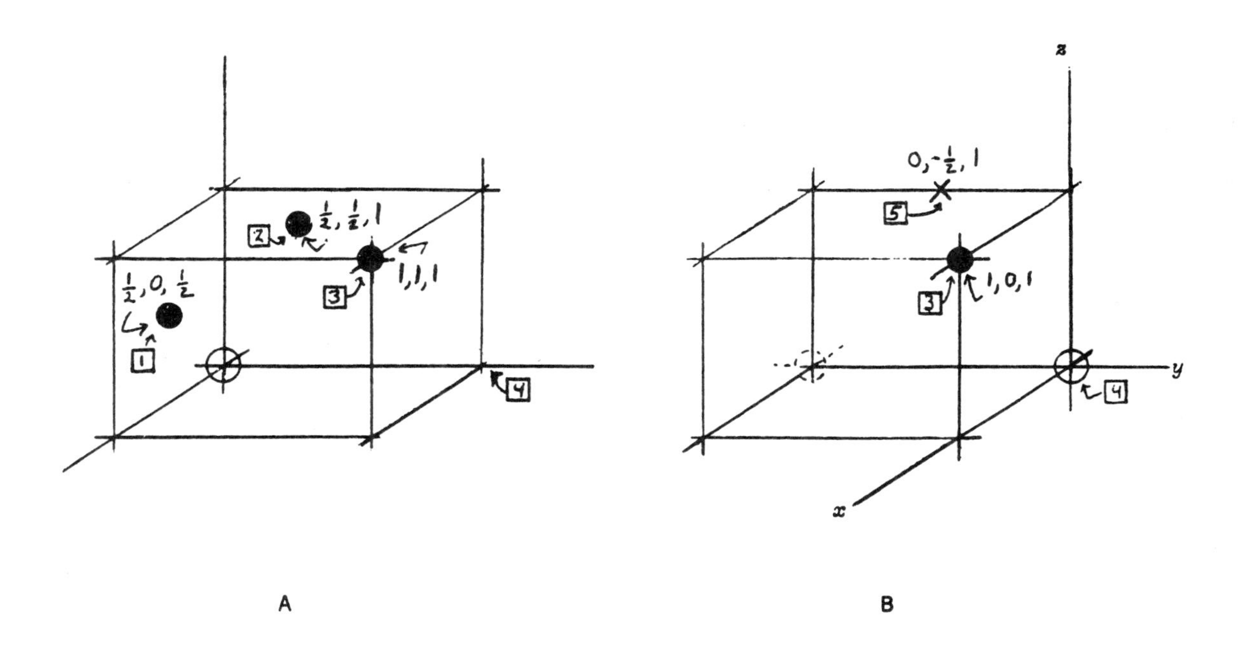

5. **Origin choice.** With conventional axes, location 1 of the previous figure is $^1/_2,0,^1/_2$; location 2 is $^1/_2,^1/_2,1$; and location 3 is 1,1,1. That is very straight forward. Now assume you had preferred location 4 as the origin (and kept the axial directions unchanged). What is the location of point 3 ? You are right if you indicated 1,0,1 (B). Likewise, point 5 is $0,-^1/_2,1$ since the y-dimension is a negative y-direction.

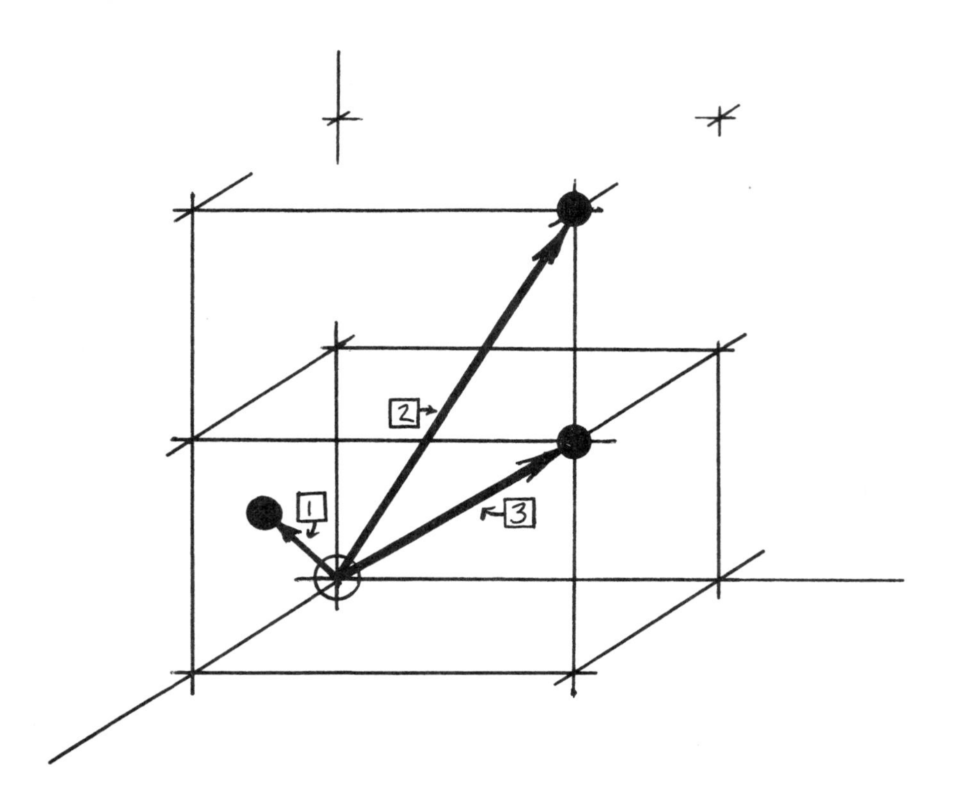

6. **Crystal directions,** [ ]. Crystal *directions* are important because properties vary with directions. *E.g.,* in iron, the relative permeability is much higher parallel to the edge of the cubic unit cell than diagonally through the cube. Conversely, deformation occurs more readily when the slip is in a diagonal direction. To index a direction, we simply start a ray at the origin and see what locations it passes through. Will you identify the directions marked 1, 2, and 3 in this sketch? Then check yourself on the next page.

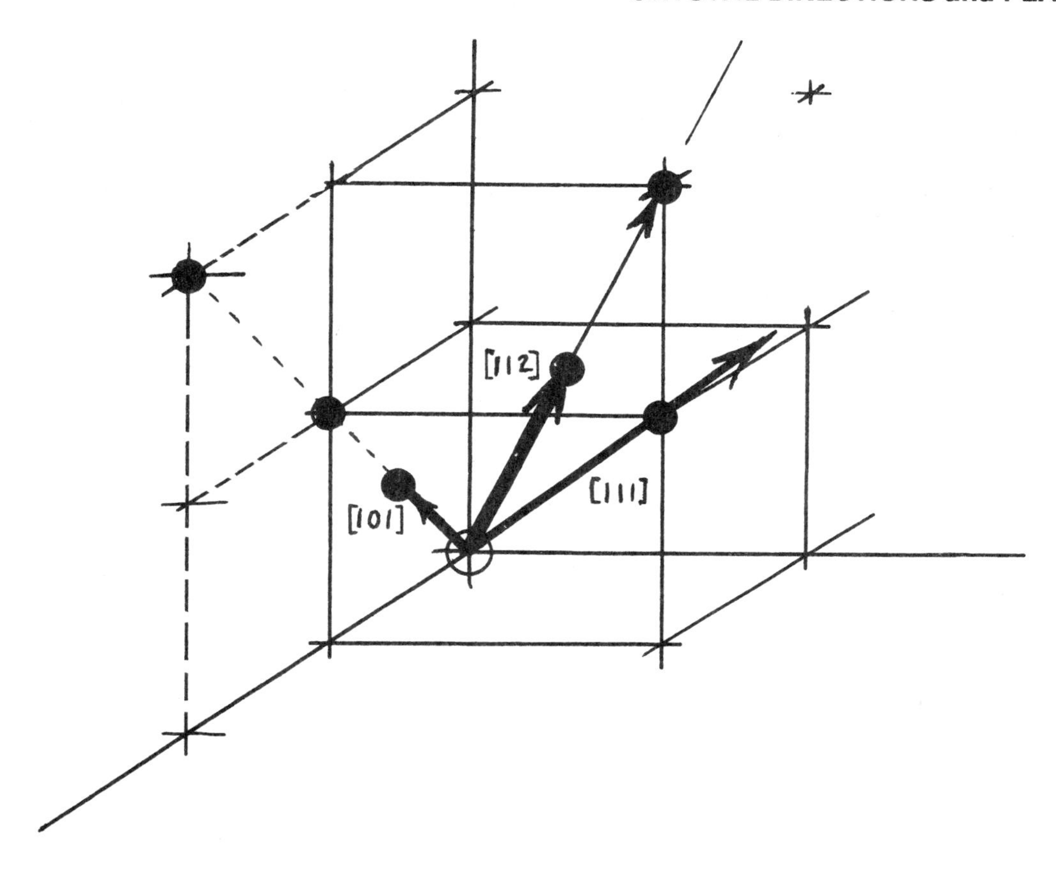

$$[\tfrac{1}{2}0\tfrac{1}{2}] = [101] = [202]$$
$$= N[101]$$

7. **Crystal directions** (con't). Consistent with earlier figures in which we identified locations, the first direction is $[^1/_2 0 ^1/_2]$; but note that the ray also extends through the location 1,0,1 so we simplify things by calling this the [101] direction. Furthermore, the [202] direction is the same; however, our *simplified rule* is that we use the index with the *lowest set of integer numbers*:

$$[^1/_2 0 ^1/_2] = [101] = [202] = . . .$$

We use [101] . The indices of the other directions sketched in this figure are [112] and [111] because they pass through 1,1,2 and 1,1,1 respectively. Of course the [112] direction also passes through the $^1/_2, ^1/_2$, 1 location (and when extended, will pass through locations 2,2,4 ; 3,3,6 ; and so on.)

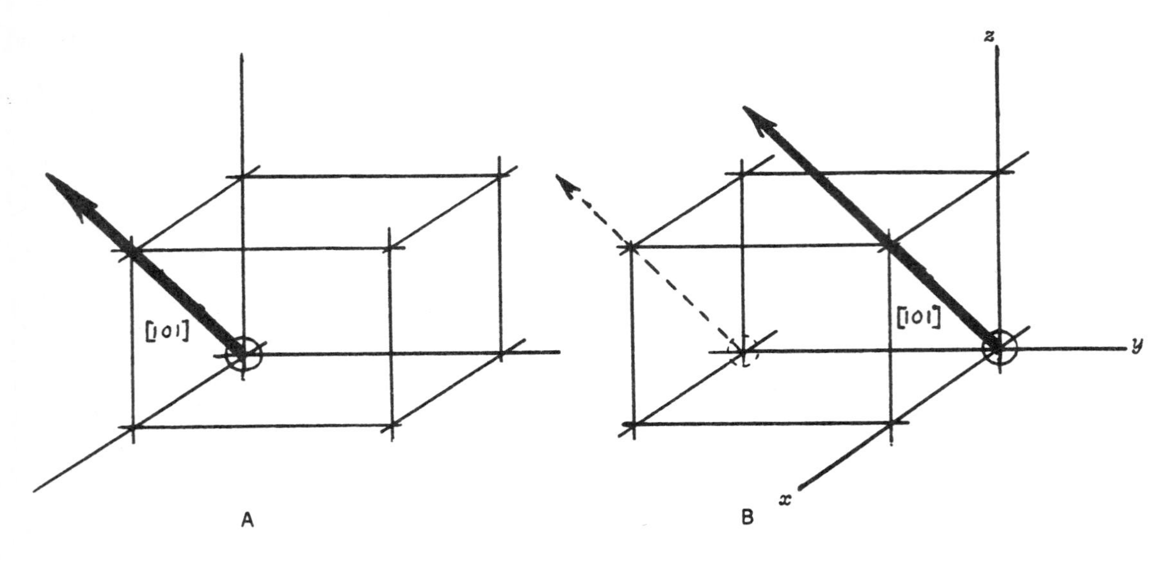

8. **Parallel directions.** Another simplifying feature is that *all parallel rays have the same index.* Let us check this out. In (A), we have sketched the [101] direction again. In part (B), we have chosen to relocate the orgin. Again, the [101] direction is identified. Although it is a different ray than in part (A), the two are parallel. Equally important, the two are identical in all respects concerning atom spacing, crystal properties, and so on. The only difference was our arbitrary choice of the origin. Hence, the above observation that *all parallel rays possess the same direction index.* Finally, note that we have enclosed our direction indices with *square brackets* [ ].

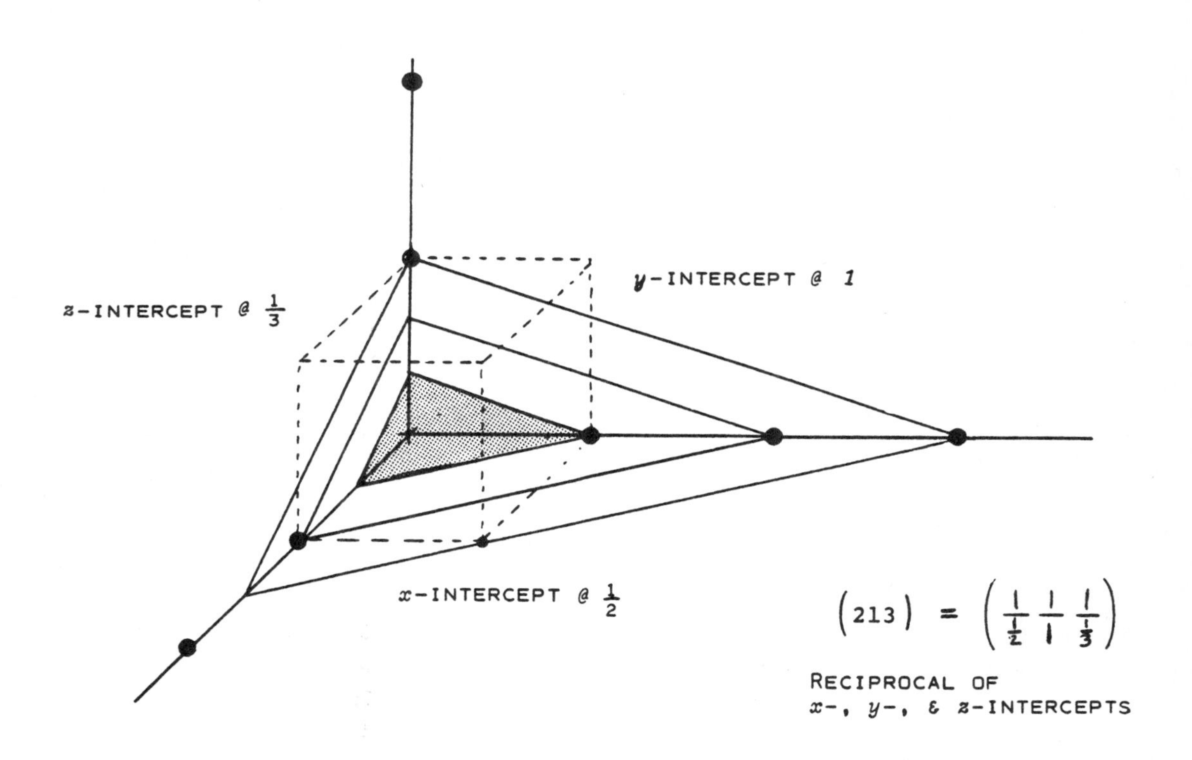

9. **Miller indices** (   ). Now let us look at planes. The planes of this figure are all *parallel*; therefore, *they will carry the same index*. Our method of obtaining the indices of planes (called *Miller indices*) is somewhat different. We do so by first identifying where the plane intercepts the three axes. The shaded plane

   intercepts the *x*-axis at $^1/_2$; its reciprocal is **2**.
   intercepts the *y*-axis at 1 ; its reciprocal is **1**.
   intercepts the *z*-axis at $^1/_3$; its reciprocal is **3**.

   This is the (213) plane.

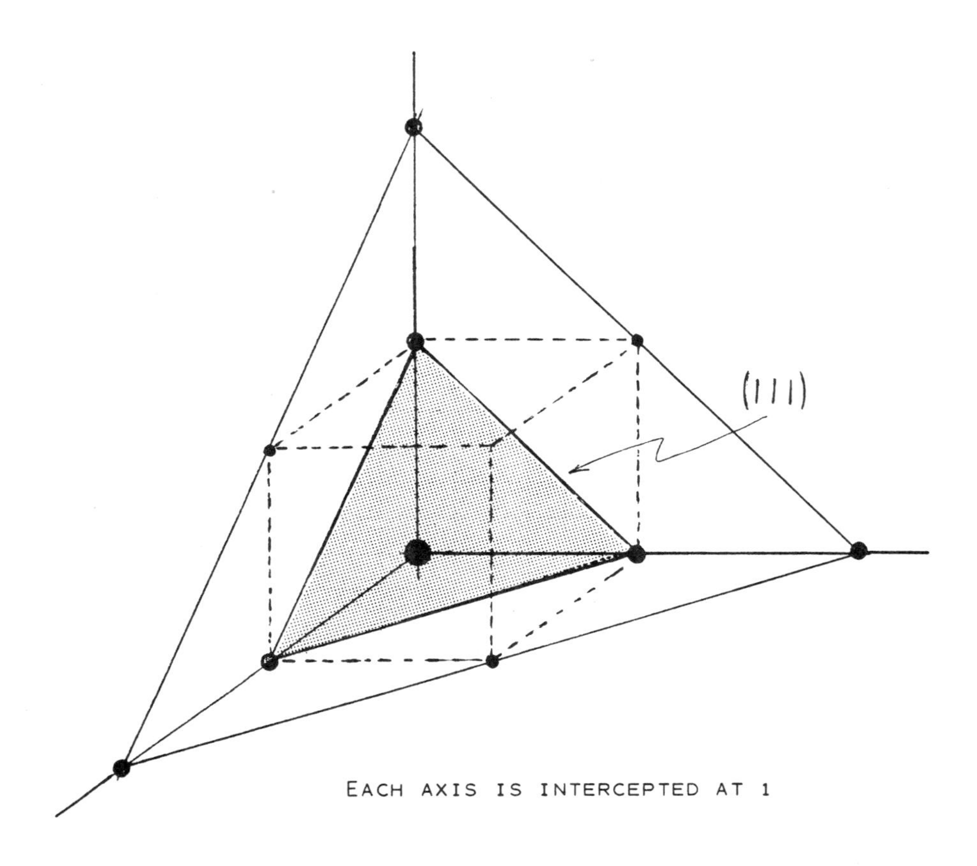

10. **(111) planes.** Actually the (213) plane of the previous sketch is not of great practical importance. Of greater importance are the (100) , the (110) , and the (111) planes. This figure shows (111) planes. Consider the shaded plane in the sketch above, which:

intercepts the *x*-axis at 1; its reciprocal is 1;
intercepts the *y*-axis at 1; its reciprocal is 1;
intercepts the *z*-axis at 1; its reciprocal is 1.

Thus, it is labeled (111). Now you sketch the (110) plane on a scrap of paper, and then check yourself with the sketch on the next page.

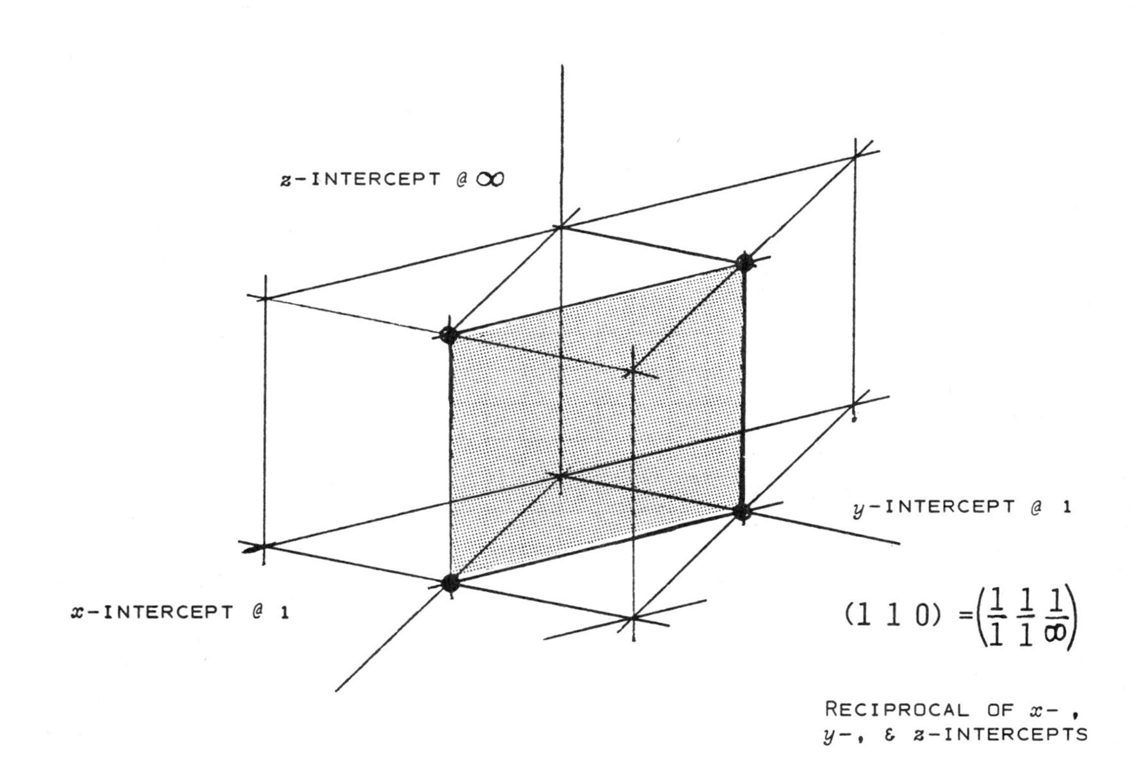

11. **(110) planes.** This is the *(111) plane,* which intercepts the *x*- and *y*-axes at unit cell distances and never intercepts the *z*-axis. Again observe that parallel planes carry the same index. Also, just as 50°C needs more than just the number 50 if we are complete with our temperature information, the numbers 110 are not sufficient. If it is a *plane*, we must use *parentheses* ( ) for closures; if it is a *direction*, we must use *square brackets* [ ] for closures; if we speak of the 1,1,0 location, we use no closure, but insert commas.

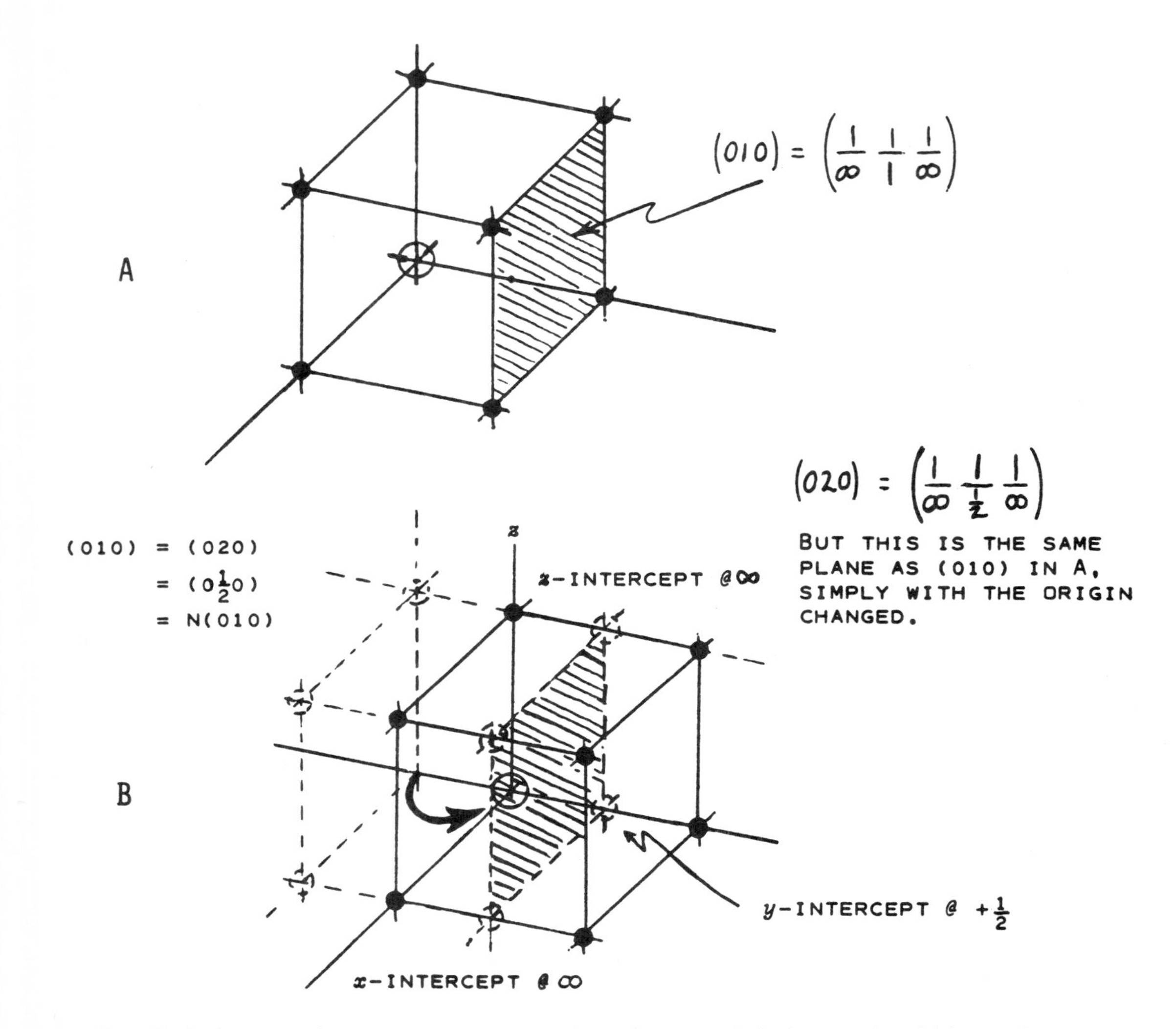

12. **Parallel planes**. Let us prove to ourselves that parallel planes should have the same notation. This is a (010) plane. We will choose arbitrarily to shift the axes to the right by one-half of the unit cell distance, (B). The original plane now:

intercepts the *x*-axis at ∞ ; its reciprocal is 0 ;
intercepts the *y*-axis at $^1/_2$; its reciprocal is 2 ;
intercepts the *z*-axis at ∞ ; its reciprocal is 0 .

We could index this plane as (020), but of course the plane has not changed; we have simply changed the reference origin. So the (010) , and the (020) labels tell us the same thing about orientation and other geometric aspects of the plane. As with directions, we can multiply the indices of a plane by any constant and not change a thing.

(010) = (020) = $(0^1/_20)$ = etc. . . .

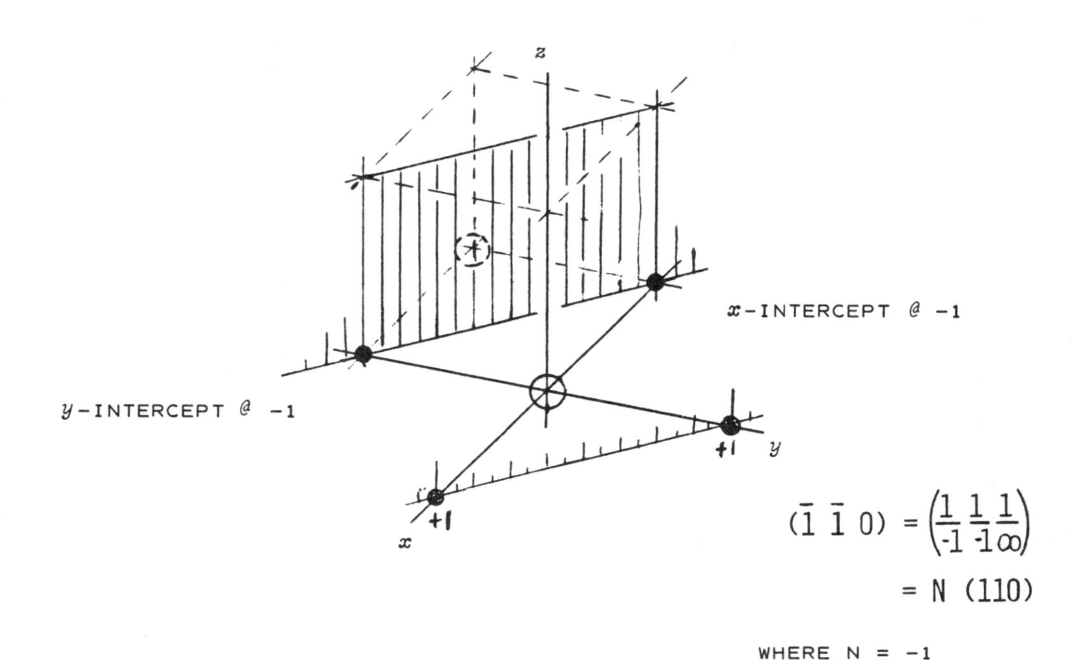

13. **Parallel planes** (con't). We can even multiply the index by a negative number. (By convention, the minus is inserted as an *overbar*).
Thus

$$(110) = (\overline{110}) = (\overline{220}) = \ . \ . \ .$$

That the $(\overline{110})$ plane is identical with the (110) plane is easily checked out in this figure by arbitrarily choosing our origins.

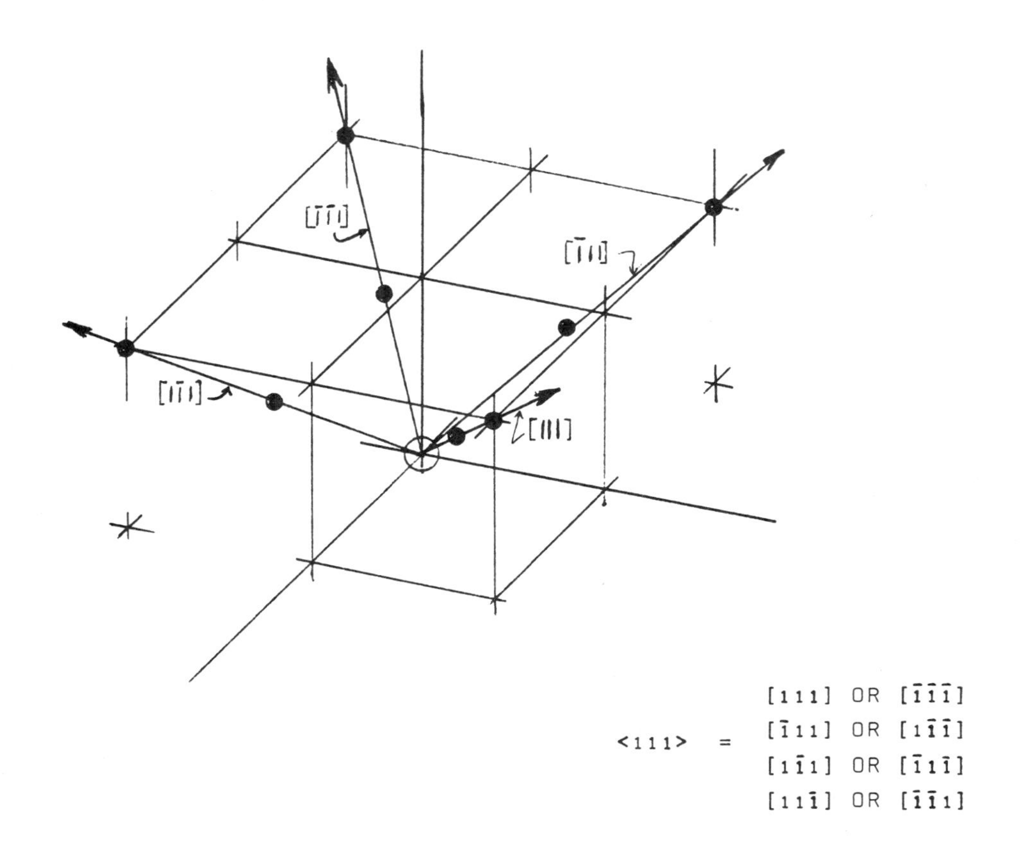

14. **Family of directions, < >.** The $[111]$, $[1\bar{1}1]$, $[\bar{1}\bar{1}1]$, and $[\bar{1}11]$ directions for a cubic crystal are shown in this figure. These directions are not parallel (a fact that we can appreciate, because $[111]$ can not be changed to $[1\bar{1}1]$ by multiplying *all* of the three indices by the *same* factor). While not parallel, they have much in common. They have the same atomic arrangements. In iron, they have the same magnetic permeability, and so on. Since they are fully equivalent, however, it is frequently desirable to cite all of these directions as a group, or *family*. When we use *pointed brackets* <111>, we mean all of the directions that are equivalent but not parallel. Thus, when we talk about deformation or magnetic permeability, we can use one family index, rather than four separate directional indices. (There are eight indices, if we count the positive and negative sense of each, *i.e.*, $[111]$ *vs* $[\bar{1}\bar{1}\bar{1}]$, which is $-1 \times [111]$.)

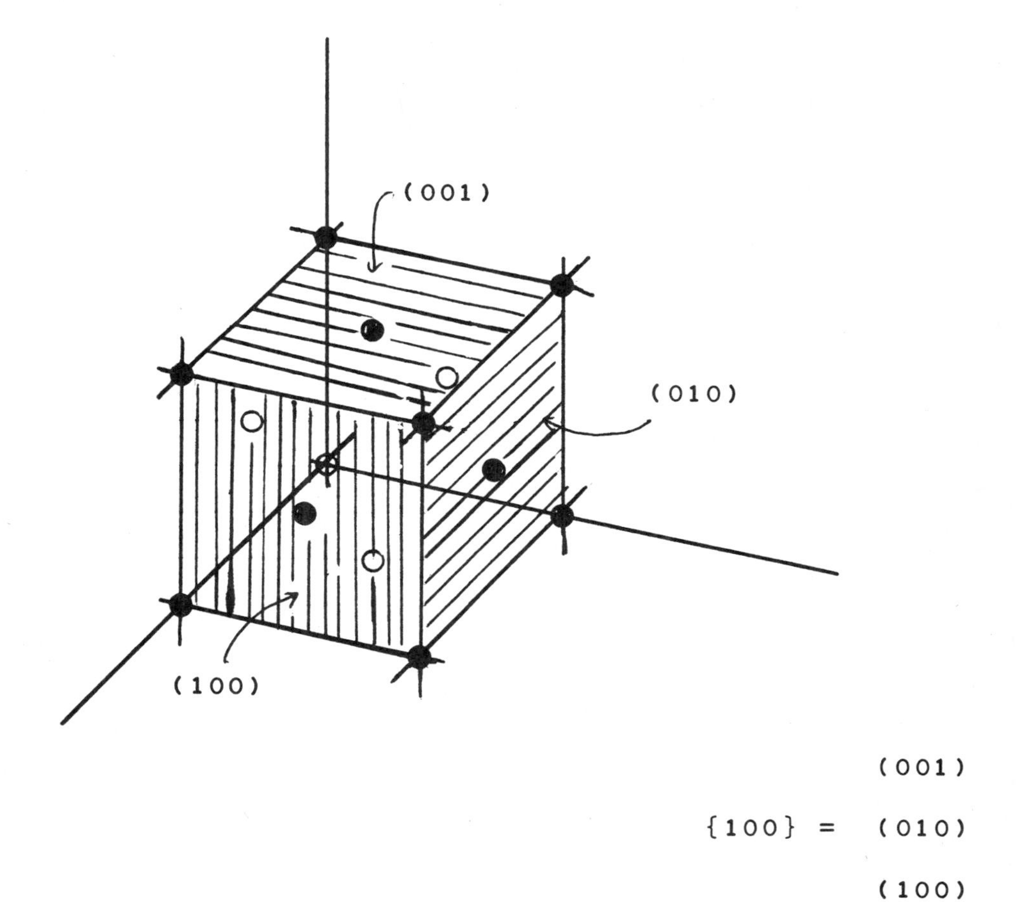

15. **Forms { }.** The (100), (010), and (001) planes are all equivalent. We can thus group these together as {100}. Note that *braces* are used rather than parentheses. We label this group of equivalent planes as a *form*, or a *family of planes*. Now sketch the {111} form on a scrap of paper; then check yourself with the next sketch.

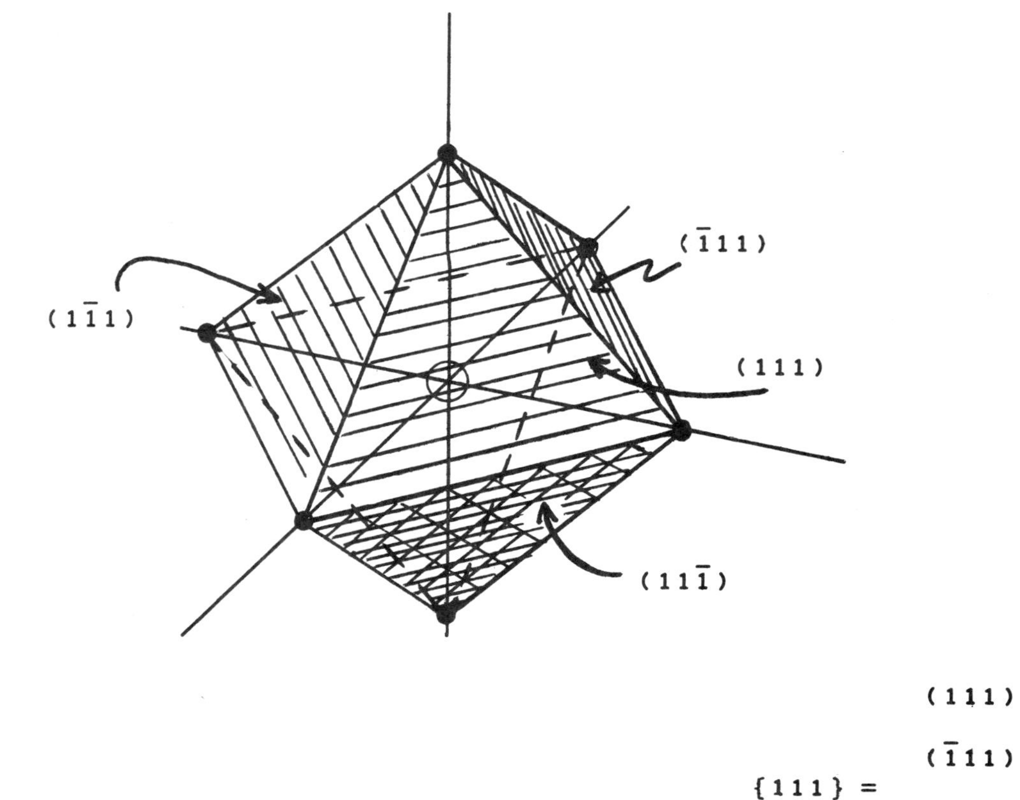

$$\{111\} = \begin{matrix} (111) \\ (\bar{1}11) \\ (1\bar{1}1) \\ (11\bar{1}) \end{matrix}$$

16. **{111} form.** These are the planes of the {111} form. In cubic crystals all are equivalent, and one could be changed to the other if we would arbitrarily choose to interchange the axes. We could also list $(\bar{1}\bar{1}\bar{1})$, $(1\bar{1}\bar{1})$, $(\bar{1}1\bar{1})$, and $(\bar{1}\bar{1}1)$, which lie on the back side of the sketch. However, since we can obtain these latter four from the four that are identified on the sketch by simply multiplying by an integer, *i.e.*, -1, we know that they form parallel pairs with (111) , $(\bar{1}11)$, $(1\bar{1}1)$, and $(11\bar{1})$, respectively.

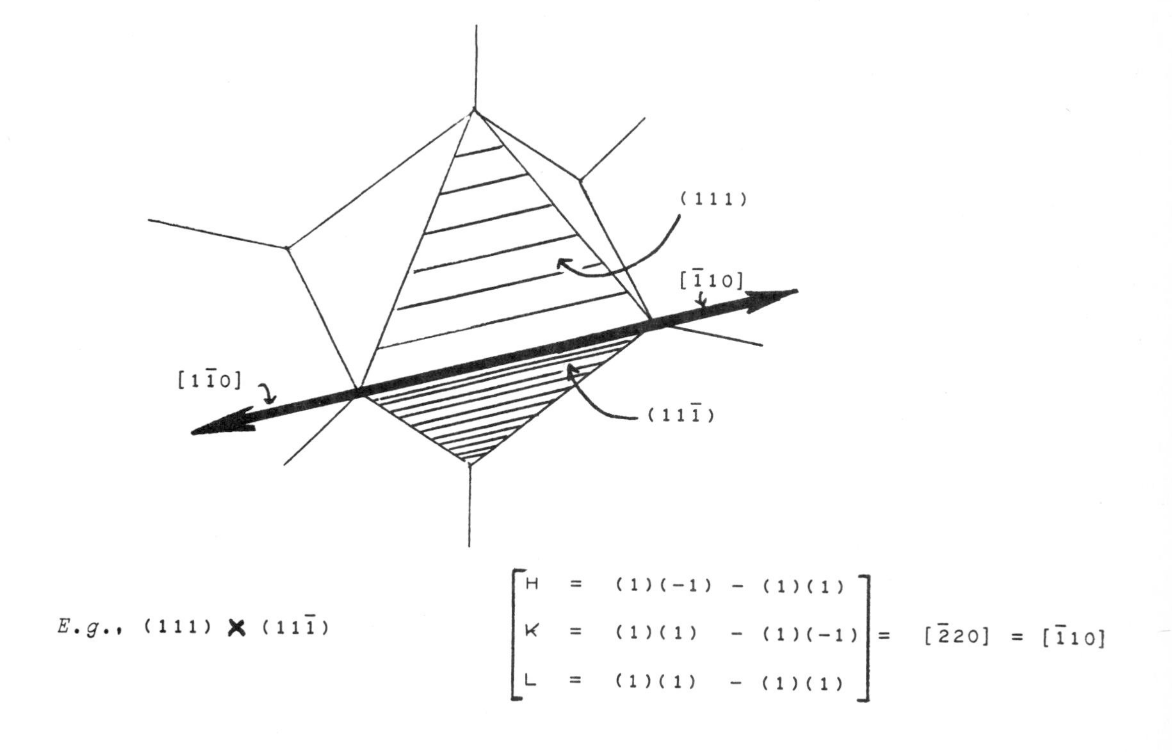

17. **Intersection of planes.** The next two sketches use the *cross* and *dot* products of the indices for planes and directions to obtain added information. In this sketch, which is nearly a repeat of the previous sketch, we can see the *line of intersection* between the (111) and the $(11\bar{1})$ planes. The *direction* of the line is $[1\bar{1}0]$, based on our earlier work, (or it is $[\bar{1}10]$ if we multiply $[1\bar{1}0]$ by -1, and point the ray in the other direction.) The index of the line of intersection of the two planes is also the *cross product* of the indices of the two planes.

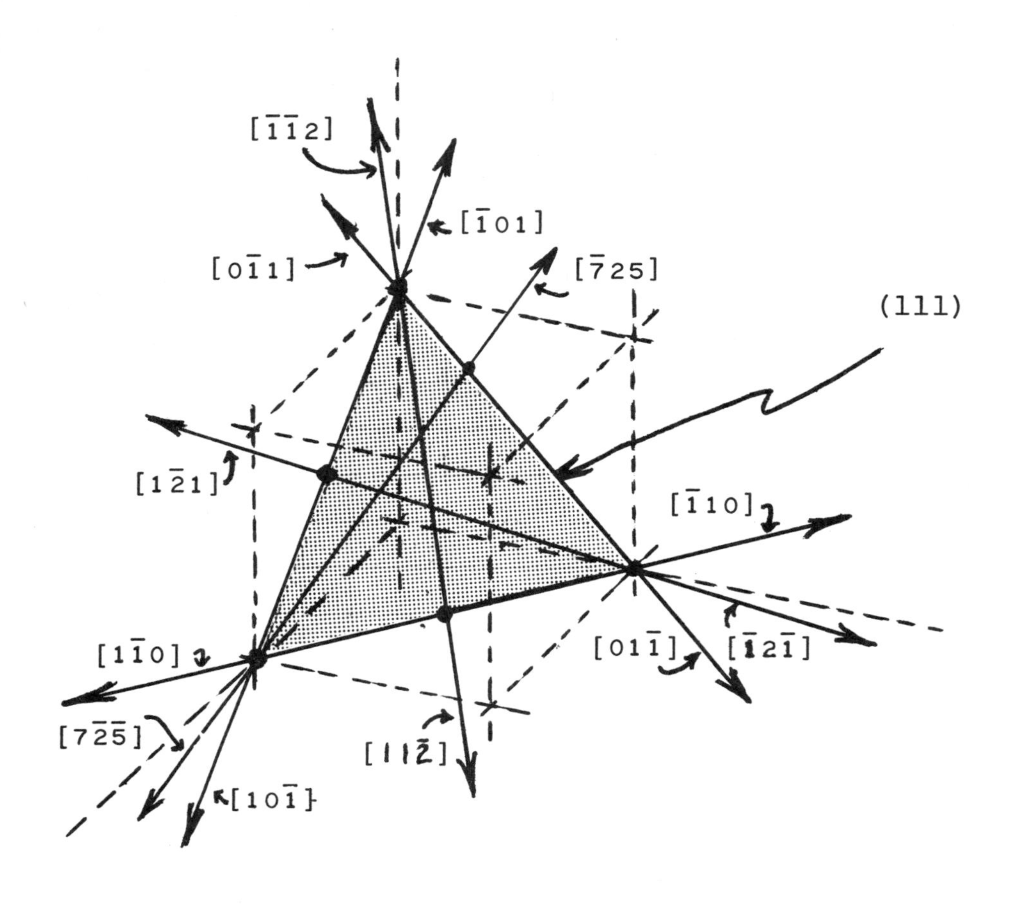

E.g., $(111) \bullet [1\bar{1}0]$ $\qquad$ $1 \bullet 1 + (1) \bullet (-1) + 1 \bullet 0 = 0$

18. **Directions within planes.** An infinite number of lines lie within one plane. From vector geometry, the *dot product* of the plane index and any direction that lies *within* that plane is zero. Thus, if we look at the above sketch, we identify a number of directions that lie in the (111) plane. Check these out, if you can calculate the dot product.

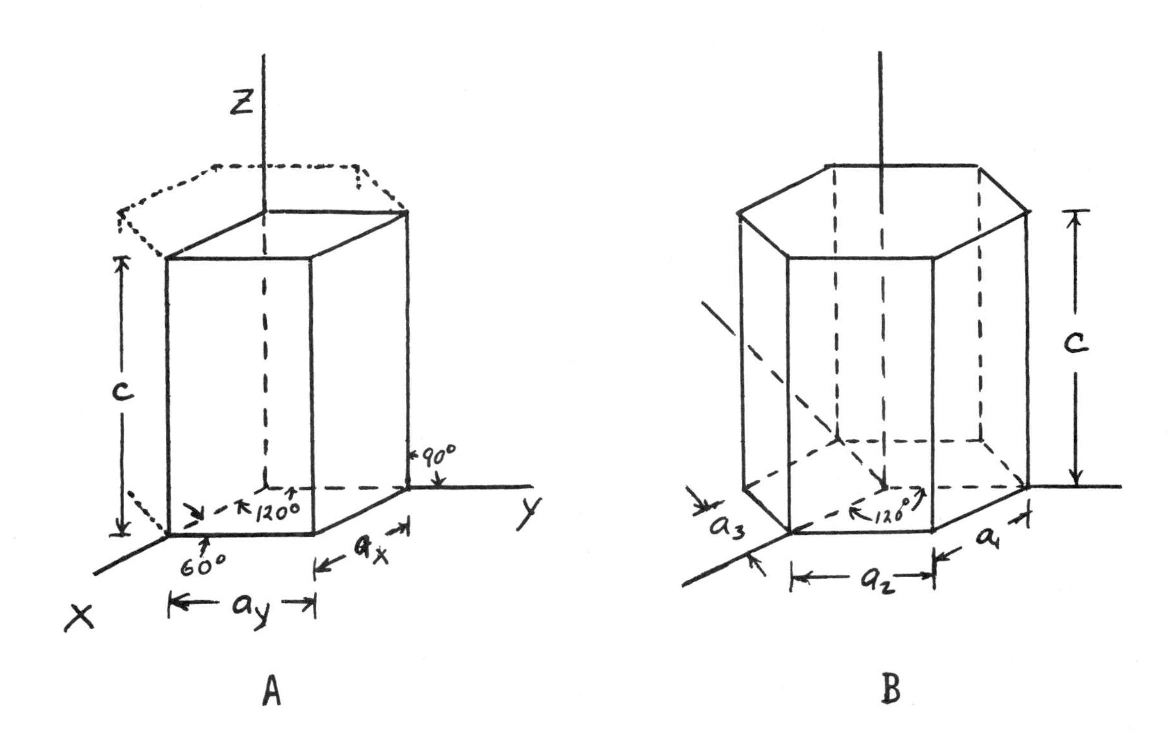

19. **Hexagonal system** (optional). The hexagonal system is shown here in two forms. In (A), it is drawn as a *rhombic* prism, *i.e.*, a prism with a rhombus base. In (B), it is a *hexagonal* prism. The geometry of the former is

$$a_x = a_y \neq c$$

Angles = 90°, 120°, (and 60°).

The geometry of the latter is

$$a_1 = a_2 = a_3 \neq c$$

Angles = 90°, and 120°.

As shown by the background sketch, the two presentations are equivalent, except for their three-to-one volume ratio.

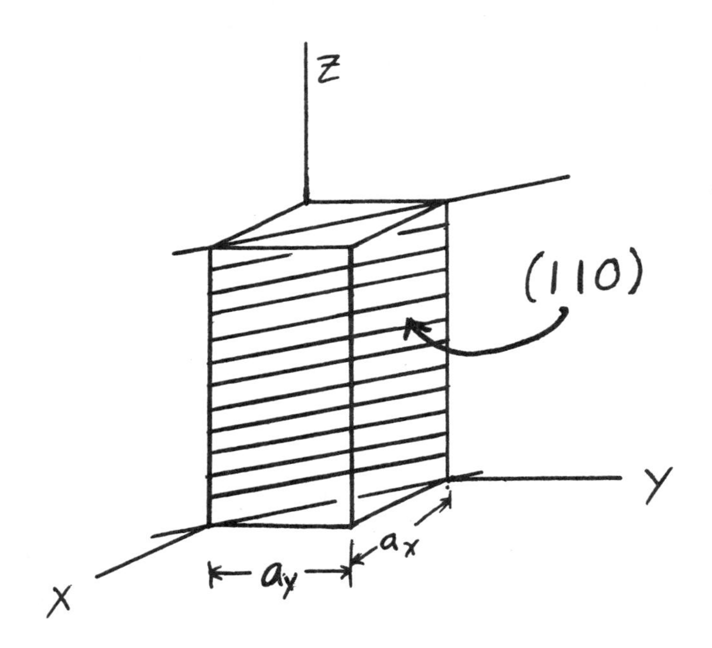

20. **Rhombic prism.** The (110) plane through a rhombic prism is shaded in this figure. As such, its Miller index is obtained identically with those of any other unit cell. We could use such indices for all of the previous crystal calculations. Sometimes, however, it is advantageous to utilize the hexagonal prism of sketch 19(B) with its four axes. Therefore, turn to the next sketch.

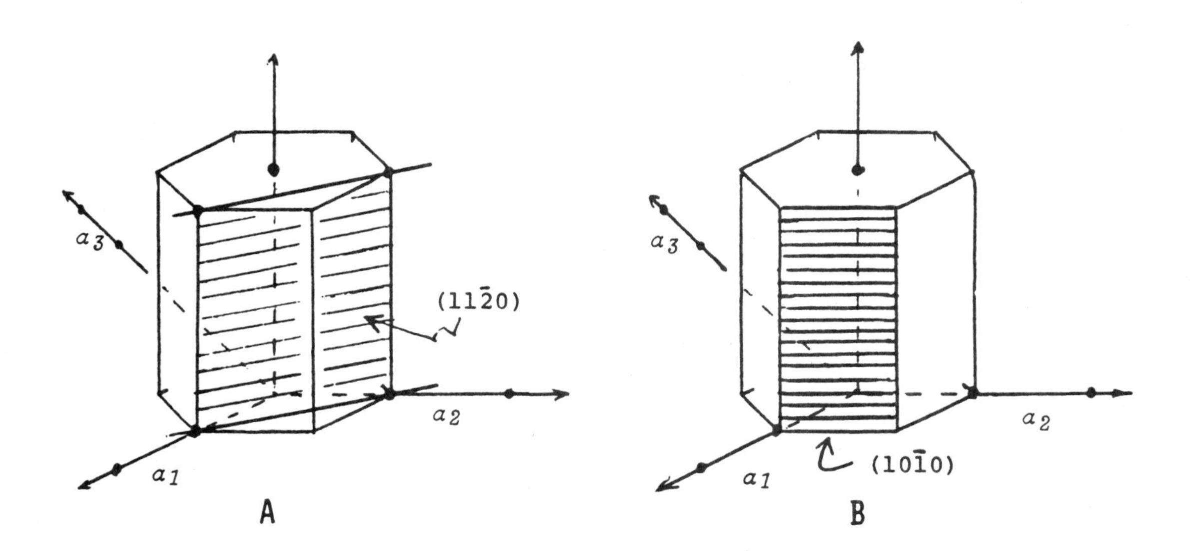

21. **Hexagonal indices**. These hexagonal sketches show three axes in the basal plane, with repeating dimensions of a_1, a_2, and a_3. [Actually the third value is redundant, because the geometry of the plane can be defined by two axes. Even so, let us use it.] The (110) plane of the previous sketch is redrawn in (A) within the hexagonal prism framework.

The shaded plane of (A) has the following intercepts and indices:

It intercepts a_1 at 1 ; its reciprocal, 1/1, is 1.
It intercepts a_2 at 1 ; its reciprocal, 1/1, is 1.
It intercepts a_3 at $-^1/_2$; its reciprocal, 1/0.5, is $\overline{2}$.
It intercepts c at ∞ ; its recioprocal, $1/\infty$, is 0.

Thus the index of the plane is (11$\overline{2}$0) , based on the four axes.

By like token, the front plane of the hexagonal prism in (B) is the (10$\overline{1}$0) plane. Thus, the {10$\overline{1}$0} form includes the six faces of the hexagonal prism.

You sketch the {10$\overline{1}$1} form on a separate sheet. Then turn to the next sketch to check your answer.

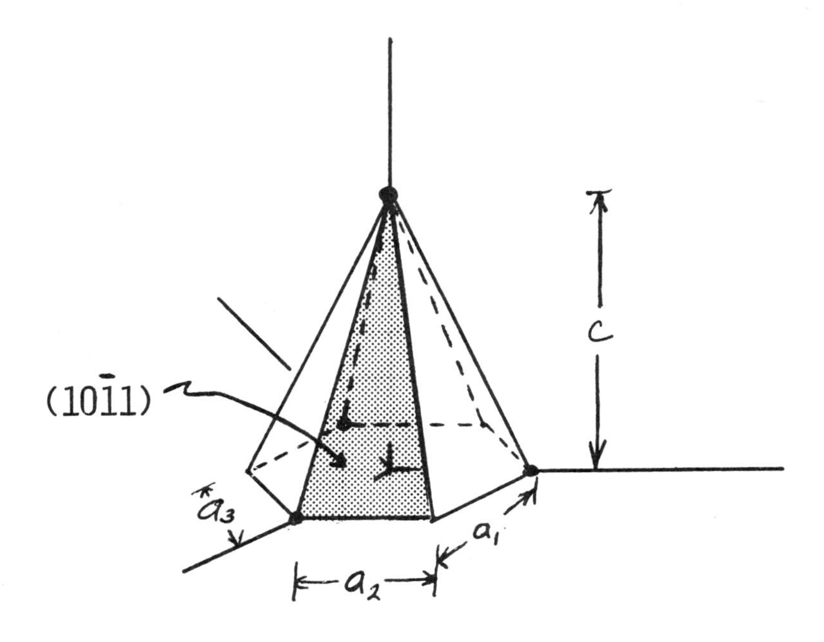

22. **Hexagonal indices** (con't). The indices of *$(10\bar{1}1)$ plane* are obtained as follows:

1 is the reciprocal of an intercept at 1 on a_1.
0 is the reciprocal of an intercept at ∞ on a_2.
$\bar{\mathbf{1}}$ is the reciprocal of an intercept at -1 on a_3.
1 is the reciprocal of an intercept at 1 on c.

This plane is shown as the frontmost plane in the above pyramid. The other *(hkil)* planes that produce the pyramid form include the a_1, a_2, and a_3 permutations of $(1\bar{1}01)$, $(0\bar{1}11)$, $(\bar{1}011)$, $(\bar{1}101)$, and $(01\bar{1}1)$. Make a point of checking out each set of indices.

RHOMBIC PRISM		HEXAGONAL
3	AXES	4
MILLER	INDICES	MILLER-BRAVAIS
(hkl)		$(hkil)$
(100)	E.G.	$(10\bar{1}0)$
		$h + k + i = 0$

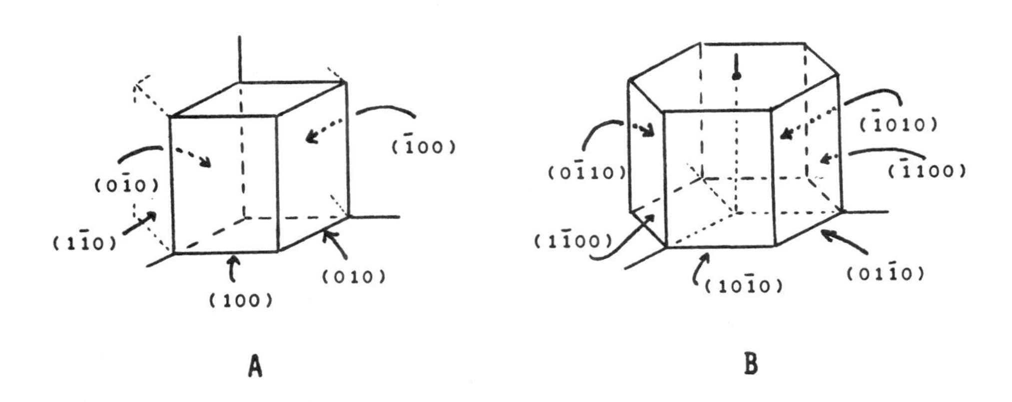

23. **Hexagonal summary .** The 4-number indices of the hexagonal system are called *Miller-Bravais* indices in contrast to *Miller* indices with three numbers. They are useful because they make the symmetry relationships of the hexagonal crystals more evident to the materials scientist and engineer. For example, the form {100} of the rhombic prism (A) does not automatically reveal that the $(1\bar{1}0)$ and $(\bar{1}10)$ hexagonal planes are identical to the (100), (010), $(\bar{1}00)$, and $(0\bar{1}0)$ planes. The form $\{10\bar{1}0\}$ leads directly to the six planes (B).

The four numbers of the Miller-Bravais indices for planes are not wholly independent, because the *sum of the first two numbers* is always the *negative of the third number.* Thus, if we label the indices as $(hkil)$; then $h + k = -i$.

Furthermore, the (hkl) of the 3-number indices are the same as the h, k, and l, of the 4-number $(hkil)$ indices. Finally, observe that the volume of the hexagonal unit cell is triple the volume of its rhombic counterpart.

ONE COMPONENT (Water)

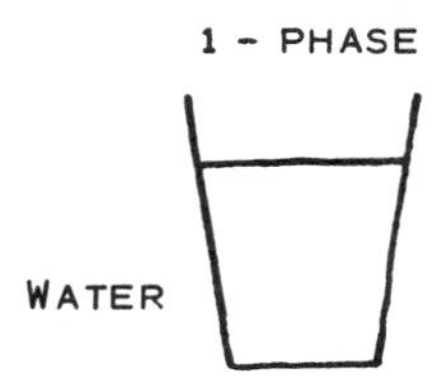

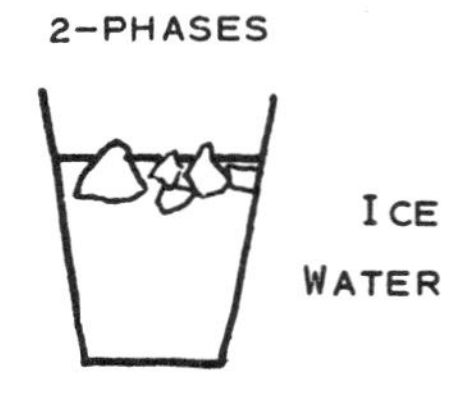

TWO COMPONENTS (Water and Sodium Chloride)

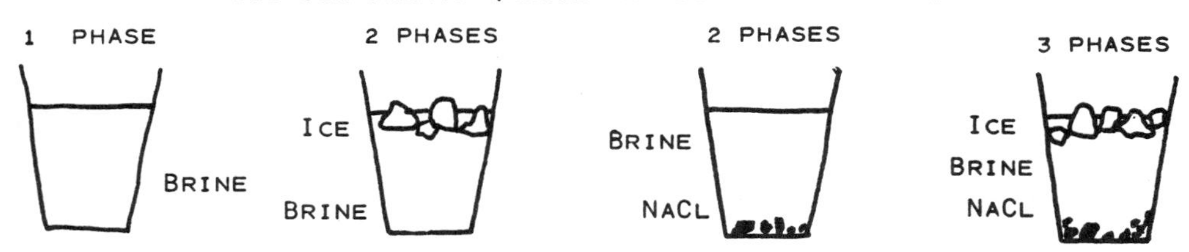

II

INTRODUCTION

to

PHASE DIAGRAMS

1. **Phases and components** (opposite). A *phase* is a homogeneous portion of matter. This means that different regions of any specific phase has essentially the same chemical composition, and the same physical properties. Examples include:

gas	liquid copper
water	solid copper
ice	α-brass

   Note particularly, that although a variety of typical metals may each have many crystalline grains, they may have only a single phase. The individual crystalline grains will differ in size, shape, and orientation, but the matter in each grain is identical. Likewise silica beach sand consists of many particles, all of the same phase, called quartz.

   The *components* of a material are the chemical species necessary to specify the chemical composition. Examples include:

   The components of a brine are H_2O and NaCl.
   The components of brass are Cu and Zn.
   The components of soft solder as Pb and Sn.

   Check your understanding of these definitions against the examples sketched above.

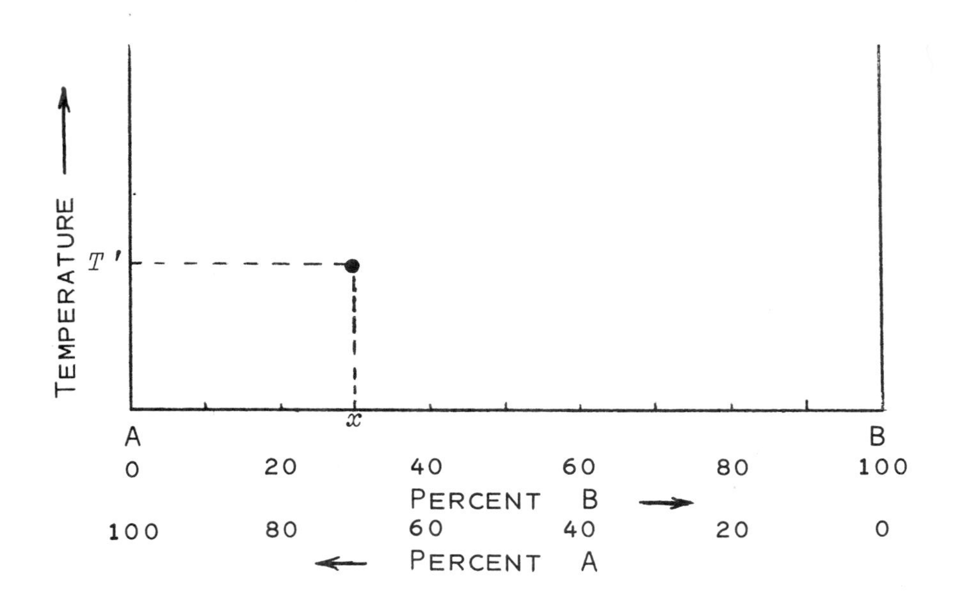

2. **Binary** (2-component) **phase diagrams.** A phase diagram is simply a plot of the equilibrated phases in materials of specified compositions and at specified temperatures.

By convention, composition is plotted horizontally and temperature vertically. Compositions are usually expressed as weight percentages, and are represented by dividing the abscissa into 100% and reading the composition of one component, B , from left to right, and the percent of the other component, A , from right to left. Obviously, for the two components, %A = 100 - %B.

A materials system of composition, x, at temperature, T', is represented by the point shown above.

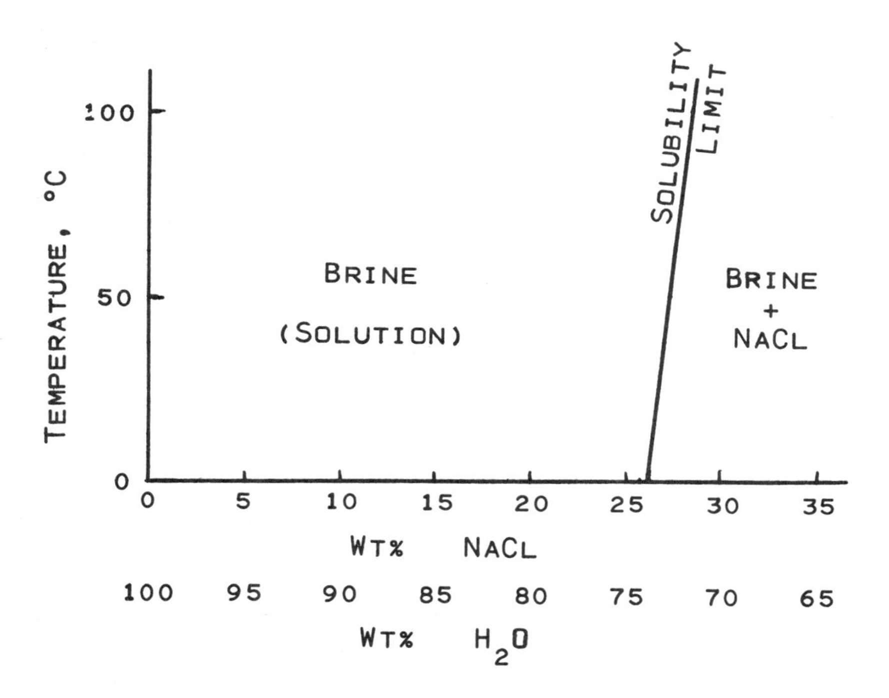

3. **EQUILIBRIUM.** Most phase diagrams can also be called *equilibrium diagrams.* They show the phases that are present in a system of a given composition when it has reached equilibrium at a given temperature . Thus, no further changes will occur in the phases that are present, no matter how much longer they are held at that temperature.

As an example, consider what happens when we add NaCl to water. At any selected temperature there is a *solubility limit.* Even though we add extra salt, no matter how long we hold a mixture of brine and solid NaCl at that temperature, we can never dissolve more salt than this limit. So we say that brine of that composition is in "equilibrium with" solid NaCl at that temperature.

Note two important points:

a) By locating any specific temperature-composition point on the above diagram, we can see what phase(s) will be present at that temperature and composition. We will use subsequent phase diagrams in an identical manner, even though they may look more complicated.

b) Most of the lines in a binary phase diagram are simply solubility limits like the one shown above.

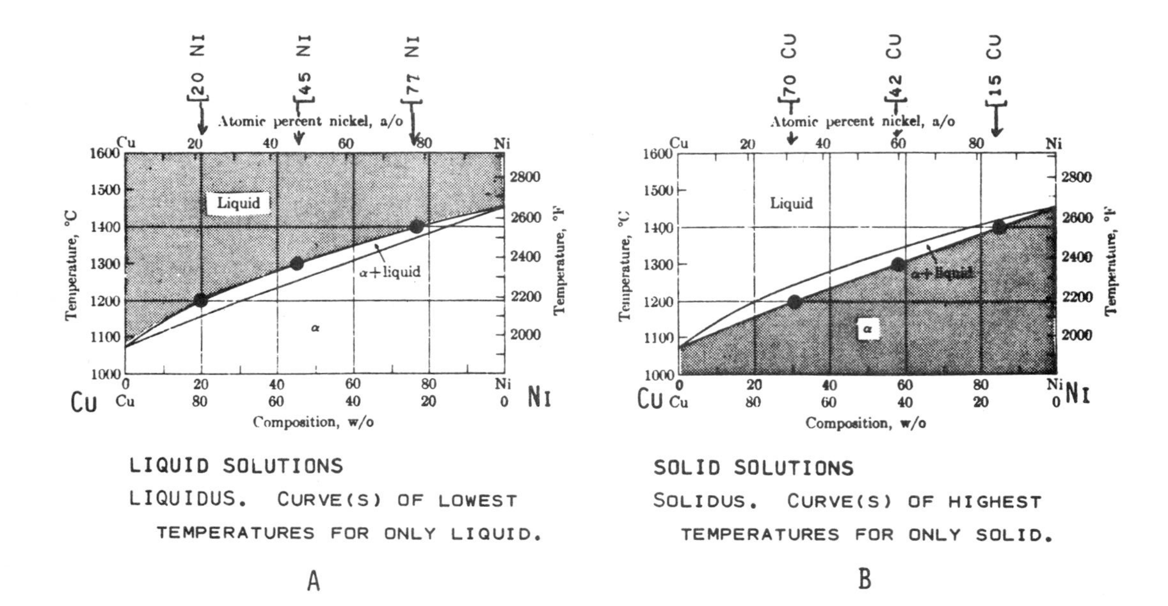

4. **The Cu-Ni system** (one phase regions). *Liquid solutions* are single phases. The shaded area in diagram (A) represents a liquid solution of nickel in copper. Its lower boundary is called the *liquidus,* because it shows the temperature above which any given alloy composition is completely liquid.

   Alternatively, this line can be regarded as the solubility limit of nickel in molten copper. At 1200°C, only 20% Ni can be dissolved. Furthermore, we see from the diagram that this solubility limit increases to 45% Ni at 1300°C, 77% Ni at 1400°C, and so on.

   Solid solutions are also single phases. The shaded area in diagram (B) is an fcc solid solution of nickel in copper (or vice versa). We call it α. The upper boundary of this solid field is called the *solidus* because all alloys below this temperature are solid. This curve can also be regarded as the solubility limit of copper in the solid. At 1200°C, 1300°C, and 1400°C, we can read that the solubility limits are 70% Cu , 42 % Cu , and 15% Cu , respectively.

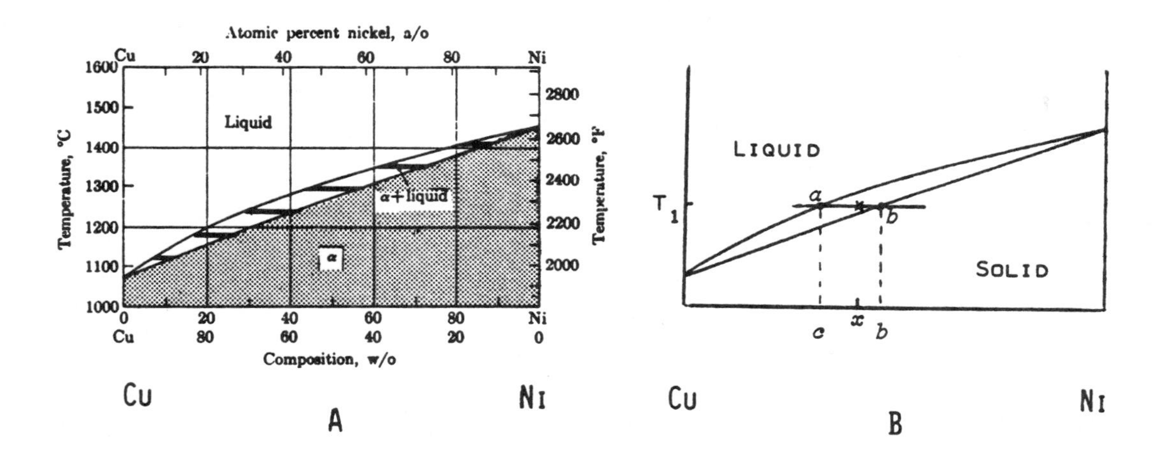

5. **The Cu-Ni system** (2-phase region). Between the liquidus and the solidus of the Cu-Ni diagram is an area where two phases must be present in equilibrium (A). It contains more nickel than is soluble in the liquid phase; it contains more copper than is soluble in the solid phase. This means there must be two phases. They are a liquid and a solid; α in this case.

   Since the liquid and the solid must be at the same temperature if they are in equilibrium, the compositions of these two phases can be read by drawing a horizontal line through the point representing the alloy, x, and temperature, T_1, of (B). The liquid has the composition, a, and the solid has the composition, b. The liquid is saturated with nickel, and the solid is saturated with copper.

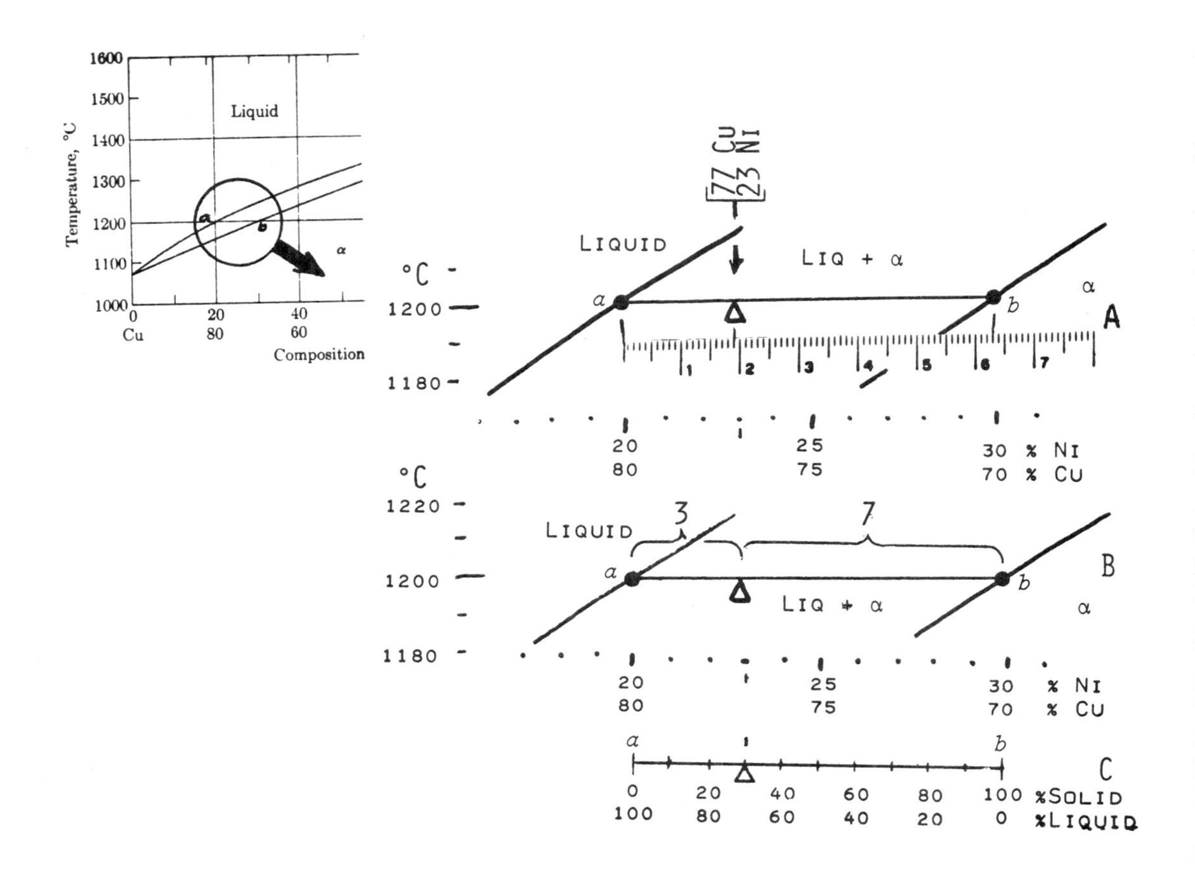

6. **Relative amounts of phases** (interpolation). In the last figure we saw how to determine the compositions when two phases are present. We now can go one step further and read the relative amounts of each of the phases.

Alloys to the left of *a* are completely liquid at 1200°C; and alloys to the right of *b* are completely solid. Between *a* and *b*, the amounts of liquid and solid will vary by a simple linear interpolation between the two extremes. An alloy midway between *a* and *b* has equal quantities of liquid and solid. We can calculate, using a scale (A), that a 77Cu-23Ni alloy at 1200°C will be 19/63, or 30%, solid and 70% liquid by weight.

Alternatively, we may use the abscissa scale on the diagram to measure the necessary lengths for the interpolation, as in sketch (B). Clearly the 77Ni-23Cu alloy is 3/10 solid and 7/10 liquid at 1200°C.

It may help to visualize this interpolation if you imagine the line $\overline{ab}$ to be divided into 100%. You then read the fraction solid from left to right, and the liquid from right to left.

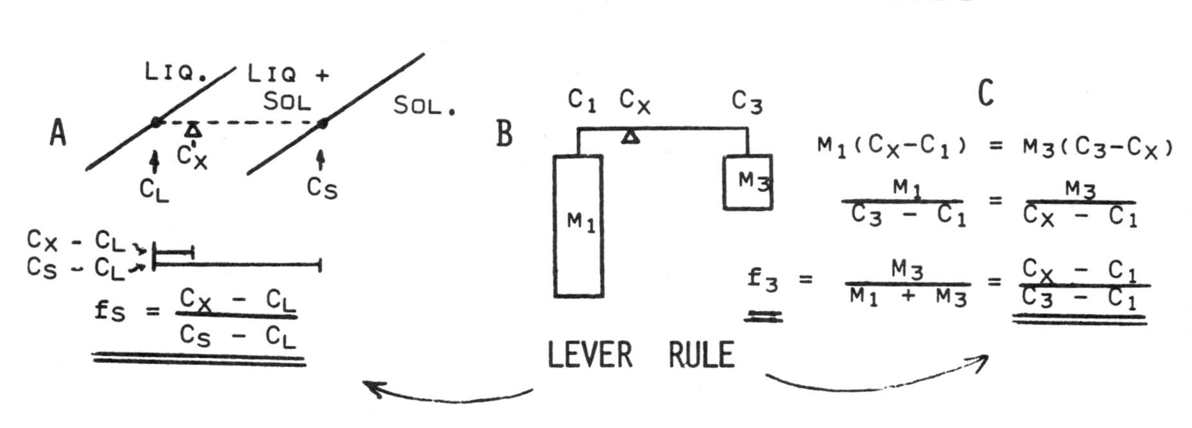

7. **The lever rule.** Our interpolation procedure of the last figure is repeated here. We generalize it by labeling the composition of the liquid, C_L, the composition of the solid, C_S, and the composition of the alloy, C_x, as shown in (A).

Our procedure for interpolation was as follows. The total distance between the two curves is $(C_S - C_L)$. The fraction of the solid, f_S, is then

$$f_S = (C_x - C_L)/(C_S - C_L) .$$

This is exactly the same equation as that derived in Physics for the equilibrium of levers. If the fulcrum in sketch (B) is at the alloy composition, the mass on the left is proportional to the lever length on the right, and the mass on the right is proportional to the lever arm on the left.

$$M_1/(C_3 - C_x) = M_3/(C_x - C_1) .$$

Therefore,

$$M_3/(M_1 + M_3) = (C_x - C_1)/(C_3 - C_1) = f_3 .$$

The equation is an *inverse ratio* rule, usually called the *lever* rule.

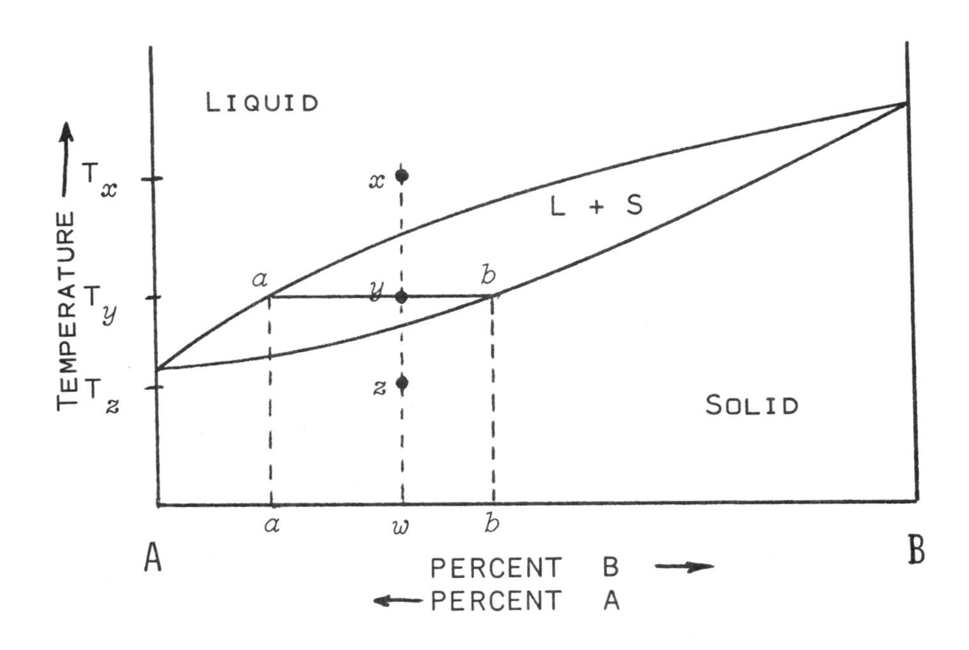

8. **Standard steps in reading any phase diagram.** From what we have learned from one of the simplest of all binary phase diagrams (Cu-Ni), we can list the steps to be taken in reading any phase diagram, no matter how complicated it may appear.
   1. Plot the point on the diagram, representing the overall composition, ***w***, and temperature, *T*, of the material.
   2. To determine what *phase*, or *phases* are present, simply read the label on the region of the diagram in which the point falls. For example, at *x* we have liquid; at *y* we have liquid plus solid.
   3. If the point falls in a single-phase region of the diagram, e.g., at *x* or at *z*, the phase must have the *same composition* as the material as a whole.
   4. If the point falls in a two-phase region of the diagram, e.g., at *y*, draw a horizontal line through it and read the *compositions of the two phases* at *a* and *b*, where this line intersects the boundaries of the region.
   5. Interpolate, using the lever rule, to determine the relative *amounts of the two phases*, e.g., fraction solid at *y* is $\overline{ay}/\overline{ab}$.

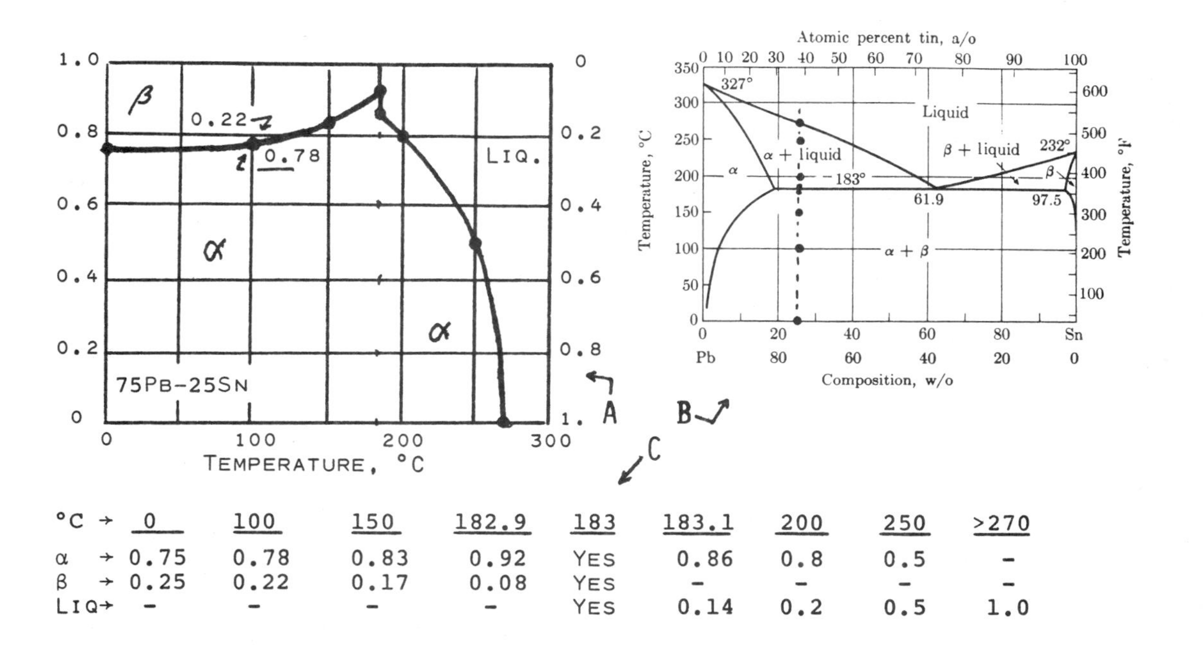

°C →	0	100	150	182.9	183	183.1	200	250	>270
α →	0.75	0.78	0.83	0.92	YES	0.86	0.8	0.5	-
β →	0.25	0.22	0.17	0.08	YES	-	-	-	-
LIQ→	-	-	-	-	YES	0.14	0.2	0.5	1.0

9. **Phase fraction charts** (optional). A *fraction chart* shows the fraction of each phase that is present, as a function of temperature, for a *specific* composition, e.g., 75Pb-25Sn. It is *not* a phase diagram. [Compare (A) and (B).] However, it is obtained from a phase diagram, simply by interpolation according to the lever rule at various temperatures as shown in (B). The results of the calculations are tabulated in (C), and plotted to give a fraction chart in (A).

That particular chart shows that, up to the eutectic temperature, the amount of α increases and the abount of β decreases with increasing temperature. Above the eutectic, the amount of α decreases and the amount of liquid increases with advancing temperatures, until 270°C and higher, the alloy is fully liquid.

An important feature can be observed at the eutectic temperature. At 183°, and only at 183°, the three phases, α, β, and liquid can exist in equilibrium simultaneously. We cannot use the lever rule when we have three phases present, but can get around the problem [as in C] by making our calculations just slightly above and slightly below the eutectic line.

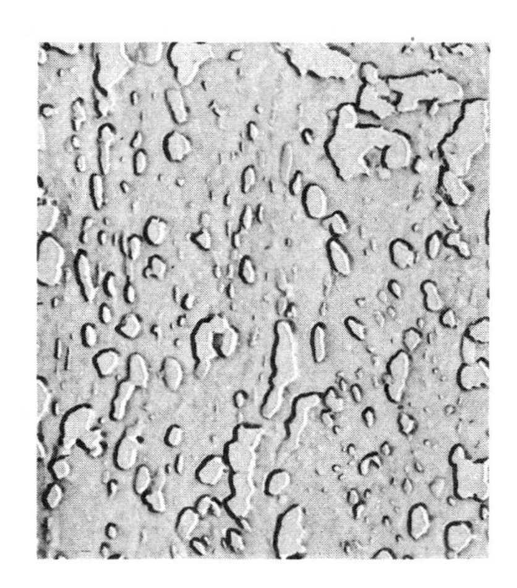 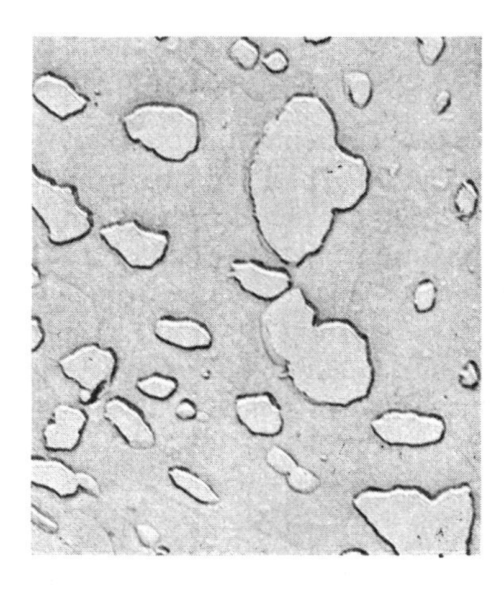

1. **Changes within solids.** Atoms within solids vibrate with thermal energy. These vibrations get more vigorous as the temperature increases. This means that a small, but significant fraction of the atoms may jump out of their crystal sites to other locations within the solid.

   Evidence of these movements is shown in the above two figures. They show the microstructure of steel that has a matrix of bcc iron. The steel contains particles of the compound, Fe_3C, or iron carbide. This microstructure was formed by heating previously quenched steel to 600°C (~1100°F). It has a hardness of 33 R_C. A change occurs after the same steel has been heated still further (675°C) for another 12 hours. The change is important, because the the hardness has dropped to only 20 R_C in the process.

III

ATOM

MOVEMENTS

1. (continued from p. 130)
   The change between these two structures requires that individual carbon atoms leave the small carbide particles and move as dissolved atoms through the bcc iron, then reprecipitate onto the larger particles. As a result, some particles are eliminated, and the average size of the remaining particles increases. There are greater distances in the soft bcc iron between the carbide particles; hence, the drop in hardness. (*Electron Microstructure of Steel*, A.S.T.M. and General Motors.)

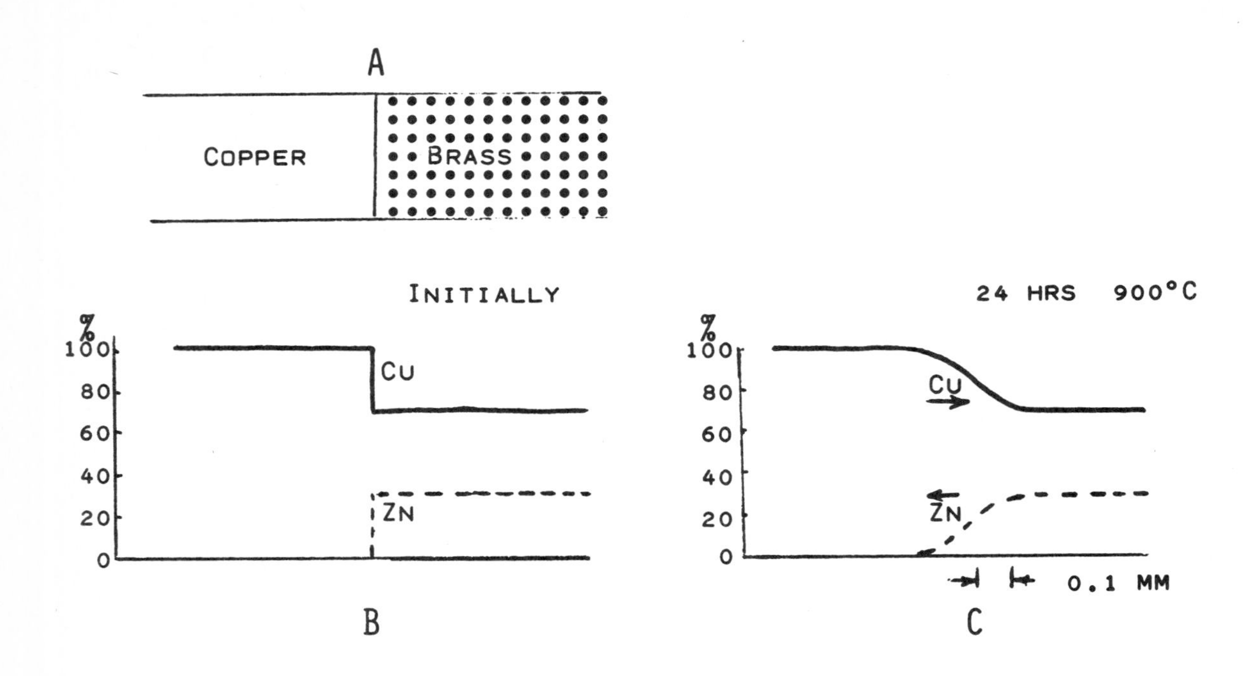

2. **Atom movements.** A second example of atom movements in solids is sketched above. Initially, a brass (a 70Cu-30Zn solid solution) made direct contact with pure copper (A). The beginning zinc and copper composition profiles are shown in (B). After one day at 900°C (~1650°F), measurements show that individual zinc atoms moved into the previously pure copper; and in addition, copper atoms have moved into the brass. Thus, the composition profile changes over a range of distance, rather than abruptly at the original interface (C).

The distances in the figure are not great; however, you can determine that 0.1 mm spans approximately 100,000 unit cells. Thus, the zinc had to move past many copper atoms in transit, and a large number of copper atoms had a net movement in the other direction.

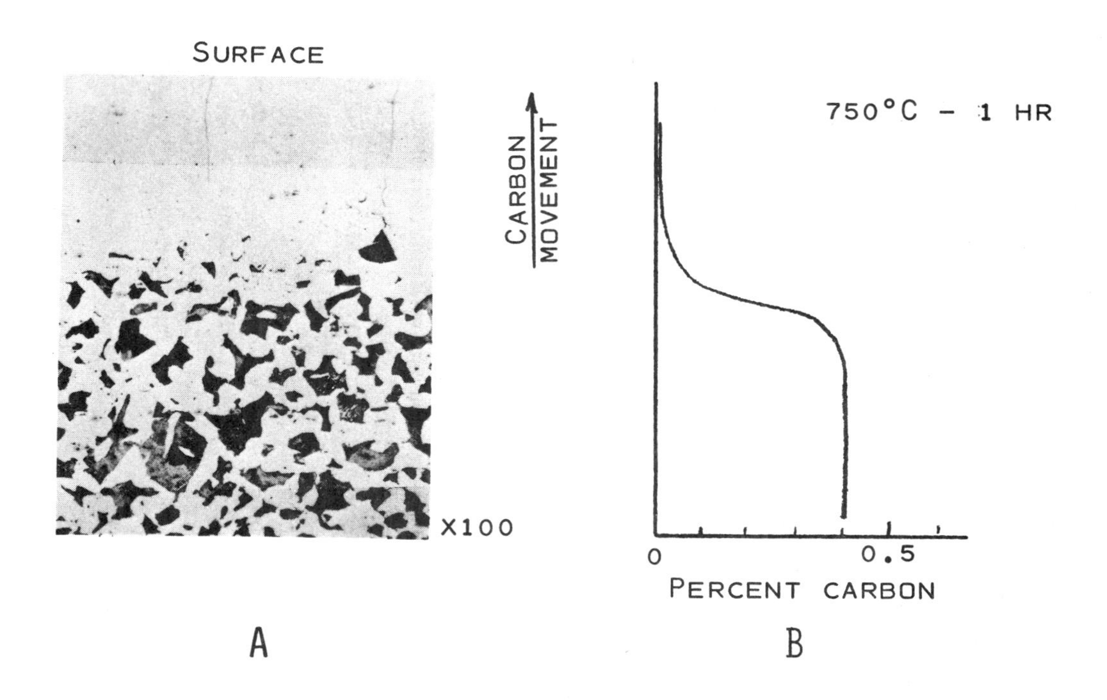

3. **Decarburization.** A third example of atom movements is shown in (A). Originally the microstructure at the bottom extended completely across the photographed area. As shown in (B), the area at the bottom is a steel with 0.4% carbon (and 99.6% iron). The area at the top has been depleted of carbon, so that it is now essentially pure iron. It has been *decarburized.*

The very top of the above sample was a surface exposed to air at 750°C (~1400°F) for an hour. The carbon in the surface unit cells reacted with the adjacent air to produce CO, which escaped as a gas. Individual carbon atoms behind the surface moved toward that surface until now there is a relatively thick carbon-depleted zone. With a depleted zone such as this, which is 0.25 mm thick, we can calculate that nearly 10^{18} carbon atoms passed through each mm^2 of surface in that hour. The initial carbon atoms were removed in a hurry, but the rate dropped off as the depleted zone thickened. However, further calculation reveals that an average of nearly 15 carbon atoms passed each unit cell each second! There can be a lot of activity within a solid.

A

$$\text{FLUX} \propto \text{CONCENTRATION GRADIENT}$$

$$J = -D(dc/dx)$$

$$\text{ATOMS}/\text{M}^2\ \text{SEC} = -D[(\text{ATOMS}/\text{M}^3)/\text{M}]$$

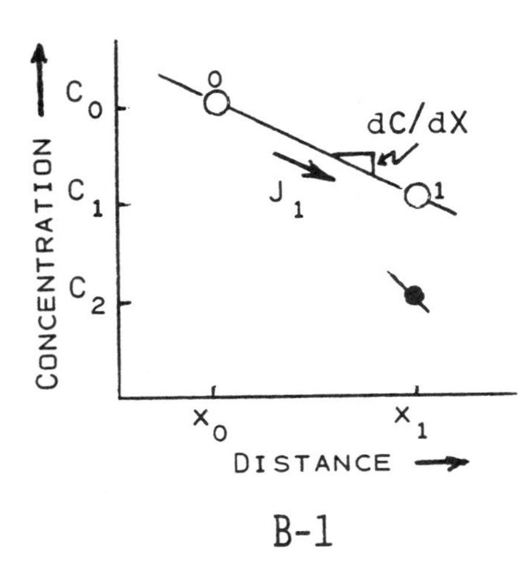

B-1

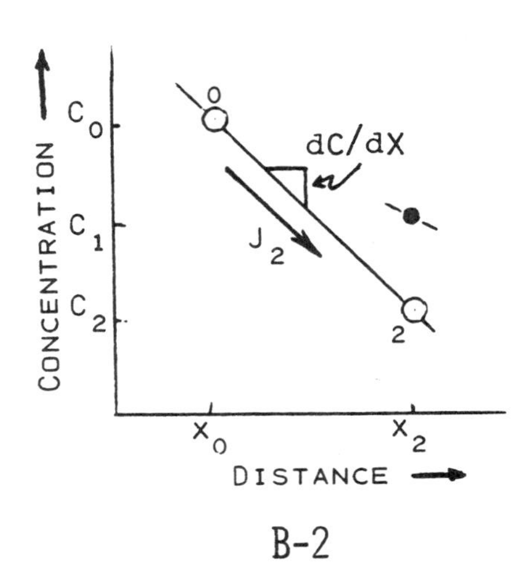

B-2

4. **Diffusion flux.** Our interest is to be able to calculate rates of atomic diffusion and to become aware of structural factors that affect those rates.

Experiments check theory in observing that the *flux*, *J*, of atoms is proportional to the concentration gradient, dC/dx. This is shown in (A) where the proportionality constant is *D*, called *diffusivity*, or *diffusion coefficient*.

We can represent this equation graphically in (B-1). Between points 0 and 1, the concentration gradient is $(C_1 - C_0)/(x_1 - x_0)$. In (B-2), we double the gradient; therefore, the flux doubles from J_1 to J_2, shown schematically by the length of the arrows. Since the flux is always positive, but the atoms move down a negative slope, the equation in (A) must include a negative sign.

Try a calculation. There is 0.3 a/o nickel in solid solution in fcc iron at point *x*. Point *y*, which is 1 mm away, has 0.2 a/o nickel within solid solution. At 1000°C, the diffusivity for nickel in fcc iron is 2×10^{-16} m^2/s. On a separate sheet of paper, calculate the flux of atoms between points *x* and *y*. Then check your answer on the next page. (Recall that an fcc unit cell of metal has four atoms. Also, since the radius of the iron atom is 0.13 nm at 1000°C, the unit cell dimension, *a*, is 0.37 nm.)

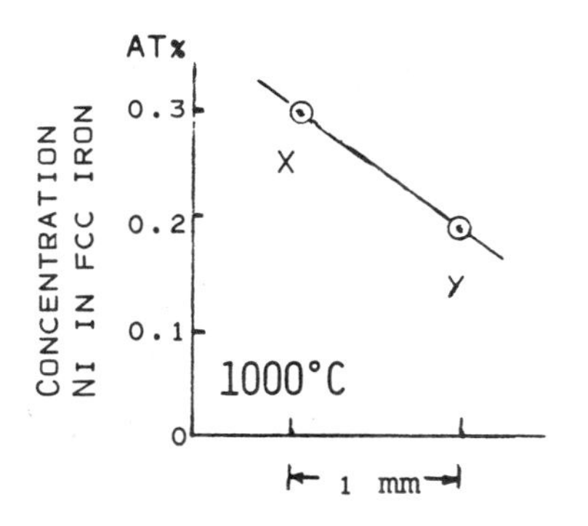

$$\text{Atoms}/\text{m}^3 = \frac{4\ \text{atoms/u.c.}}{(0.37\times 10^{-9}\ \text{m})^3/\text{u.c.}} = 8\times 10^{28}\ \text{atoms}/\text{m}^3$$

$$\text{Atoms of Ni at Y} = \left(0.002\frac{\text{Ni atoms}}{\text{atom}}\right)\left(8\times 10^{28}\frac{\text{atoms}}{\text{m}^3}\right) = 1.6\times 10^{26}\frac{\text{Ni atoms}}{\text{m}^3}$$

$$\text{Atoms}\ \text{"}\ \ X = (0.003\ \text{"}\ )(\ \text{"}\ ) = 2.4\times 10^{26}\ \text{"}$$

$$\text{Conc. gradient} = (1.6 - 2.4)(10^{26}\text{Ni}/\text{m}^3)/(0.001\ \text{m}) = -8\times 10^{28}\text{Ni atoms}/\text{m}^4.$$

$$\text{Flux} = -(2\times 10^{-16}\ \text{m}^2/\text{sec})(-8\times 10^{28}\text{Ni atoms}/\text{m}^4)$$

$$\longrightarrow\quad J_{1000C} = \underline{1.6\times 10^{13}\text{Ni atoms}/\text{m}^2\cdot\text{sec}}.$$

$$\longrightarrow\quad J_{750C} = -(3\times 10^{-19}\ \text{m}^2/\text{sec})(-8\times 10^{28}\text{Ni atoms}/\text{m}^4)$$

$$= \underline{2.4\times 10^{10}\ \text{Ni atoms}/\text{m}^2\cdot\text{sec}}.$$

4. **Diffusion flux calculation** (con't). With four atoms per unit cell, which has a volume of $(0.37 \times 10^{-9}\ \text{m})^3$, there are 4 atoms/$(50 \times 10^{-30}\ \text{m}^3)$, or 8×10^{28} atoms/ m^3.

   A calculation shows that the concentration gradient between *x* and *y* is -8×10^{20}Ni atoms/m^4. This gives a flux of 1.6×10^{13} Ni atoms/$\text{m}^2\cdot$s at 1000°C, since $D_{1000°C}$ is $2 \times 10^{-16}\text{m}^2$/s.

5. **Another calculation and units.** At 750°C, the diffusion coefficient is $3 \times 10^{-19}\text{m}^2$/s. Thus, the previous gradient produces a flux of only 2.4×10^{10} Ni atoms/$\text{m}^2\cdot$s , down by a factor of ~700.

   Let's review our units at this point. *Flux* is in atoms/$\text{m}^2\cdot$s ; a *concentration gradient* is in (atoms/m^3)/m, or atoms/m^4. Therefore, the *diffusivity* must be in m^2/s. We will see related units when we talk about electron mobility in semiconductors where the units are (m^2/s) per volt; or in fluid flow, where the units for fluidity are (m^2/s) per unit force.

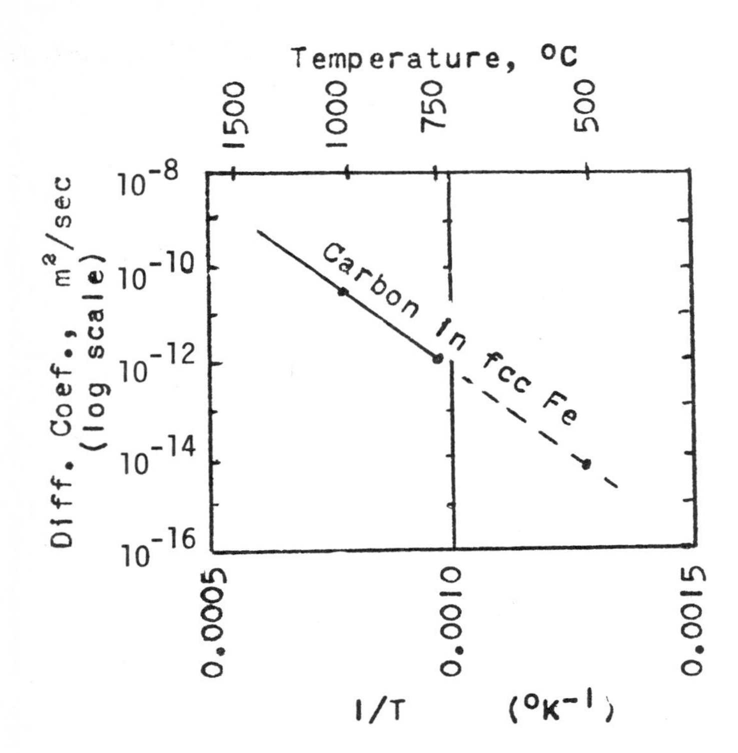

LOG D = A - B/T

Y = A - BX

6. **Diffusion coefficient** (or diffusivity). The value of D now attracts our attention, because it is the variable that dictates the flux for a given gradient. Its value depends on a number of factors. We have just seen that *diffusivity* is lower at lower temperatures. The variation of D with temperature, T, is more complex than the flux-concentration gradient relationship we have just calculated. As shown here:

$$\log D = A - B/T.$$

This is comparable to the algebraic equation,

$$y = a - bx,$$

where y is log D and x is $(1/T)$. (T is expressed in kelvin, K.) Thus, we have the plot shown in the graph for the value of D for carbon diffusion in fcc iron. The feature which may be unnatural to us is that the abscissa is *reciprocal temperature,* so that the more elevated temperatures are to the left (top abscissa).

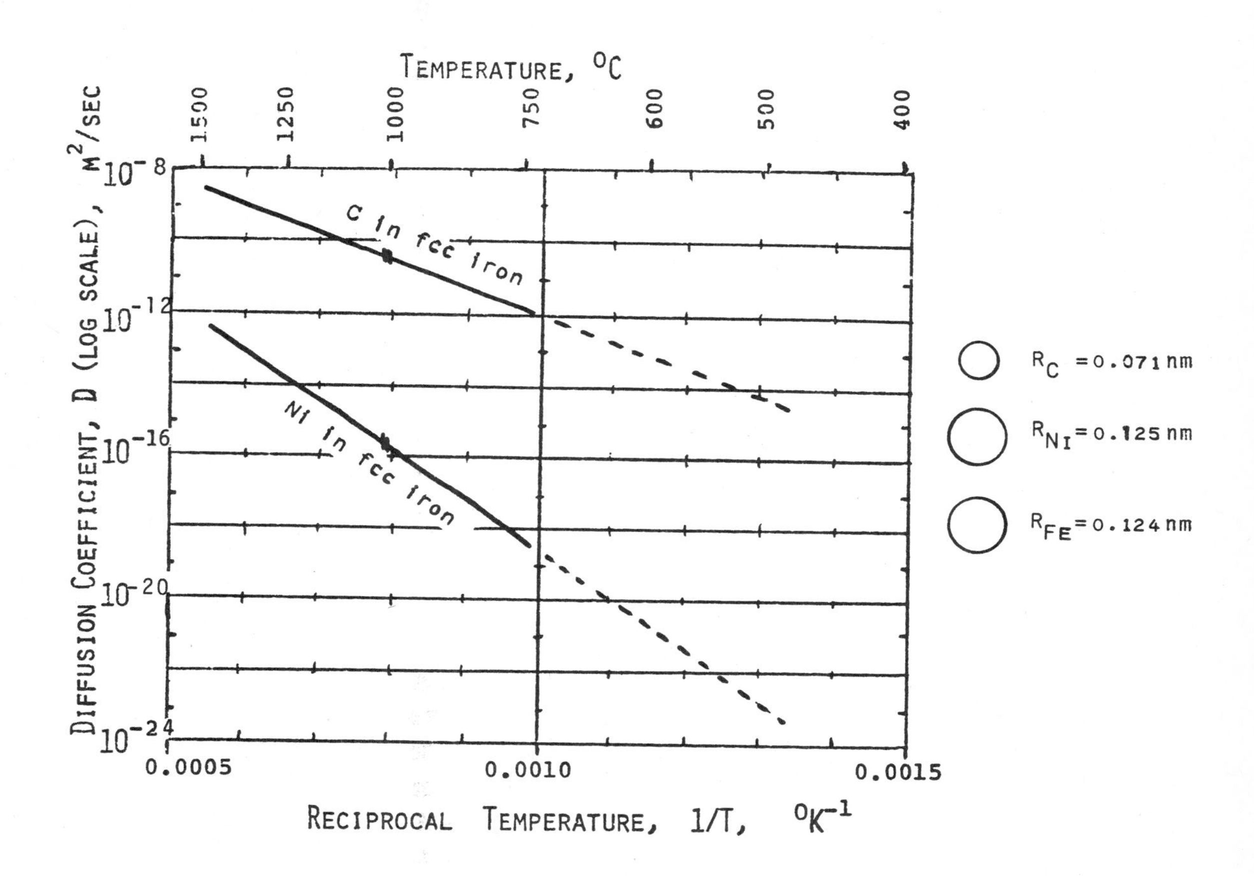

7. **Diffusivity *vs.* atom size.** This figure shows the values of the diffusion coefficient of carbon in fcc iron and of nickel in fcc iron. The difference is pronounced,-- about five orders of magnitude at 1000°C, and more at lower temperatures.

This difference is easy to explain. The carbon atom has a radius that is only 60% as great as that of the nickel atoms and the iron atoms. Therefore the carbon atom can move among the iron atoms *interstitially*. The larger nickel atoms have to move among the iron atoms by diffusion in a *substitutional* solid solution.

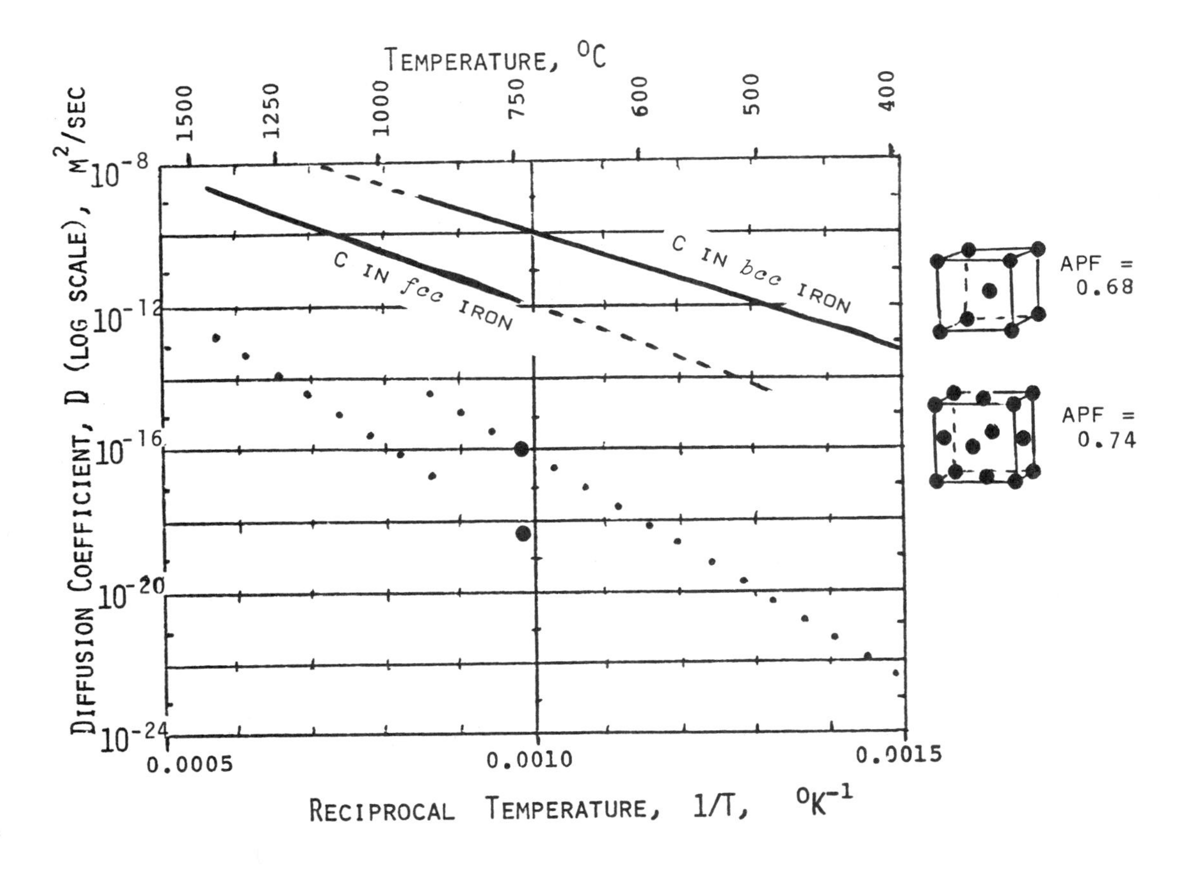

8. **Diffusion coefficient *vs.* atomic packing.** Here is another comparison. The diffusivities of carbon in bcc iron are greater than of carbon in fcc iron. (Recall that iron is bcc at low temperatures and fcc at higher temperatures. With appropriate procedures, we can have either, or both, in the 750-900°C temperature range.)

At 750°C, $D_{fcc} = 10^{-12}$ m^2/s, and $D_{bcc} = 10^{-10}$ m^2/s, differing by a factor of 100.

Again, we can understand this difference because the bcc structure is much more open than the fcc structure. The *atomic packing factor* of bcc metals is 0.68; of fcc metals is 0.74. The increase in open space from 26 v/o (fcc) to 32 v/o (bcc) permits the atoms to move around more readily; in fact, at 750°C, they can diffuse 100 times as readily. Not only are the channels among the atoms slightly larger and more numerous, it is much easier for the iron atoms to be shoved aside in the looser structure to make way for the diffusing atoms.

The dotted curves in the above sketch are for diffusion coefficients of iron atoms in fcc iron and iron atoms in bcc iron. They will be explained in the next figure. However, before you look ahead, can you identify which curve is which ?

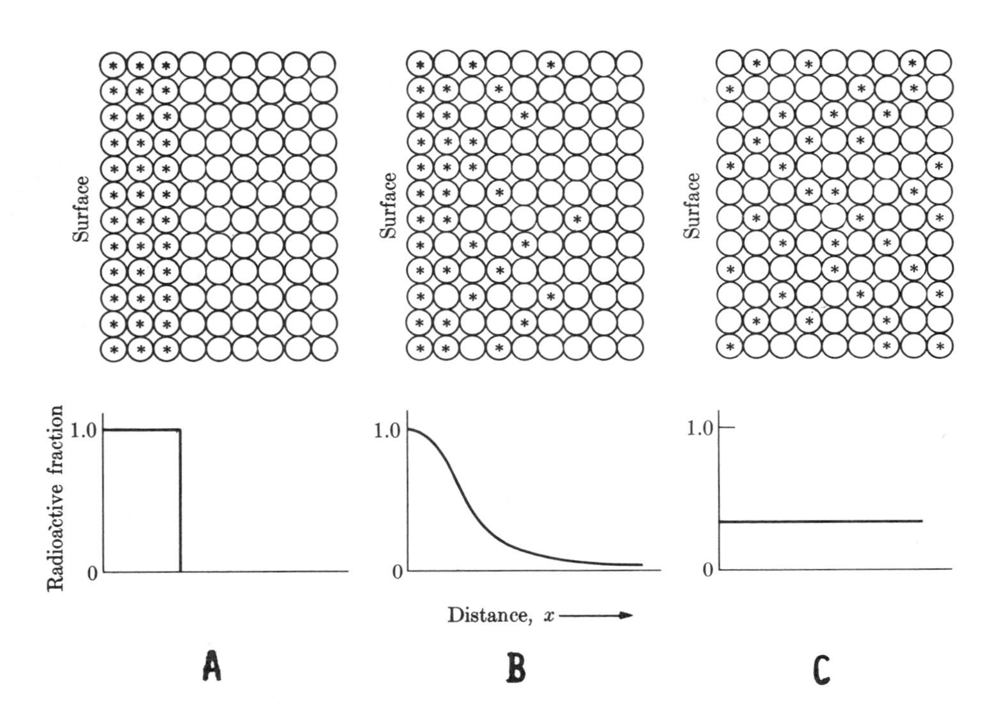

9. **Self diffusion.** *Self diffusion* indicates that iron atoms will move about in an iron crystal; likewise, nickel atoms will rearrange themselves in nickel crystals; and so on. We can not detect this by chemical analyses, because the host structure and the moving atoms have the same composition. We can, however, detect self diffusion by radioactive tracers as shown in this figure, where $time_C > time_B > time_A = 0$.

   Assume that this is nickel onto which a radioactive isotope has been electroplated. Because of its radioactivity, we can detect where those isotopes are. Initially, they are all on the external surface (left). With time, however, these atoms spread into the interior. By measuring the distribution of these isotopes after succeeding periods of time, it is possible to calculate the diffusion coefficient for self diffusion. For iron in bcc iron, $D = 10^{-16}\ m^2/s$ at 750°C; for iron in fcc iron, $D = \sim 3 \times 10^{-19}\ m^2/s$. These data are shown as dotted curves in the previous sketch.

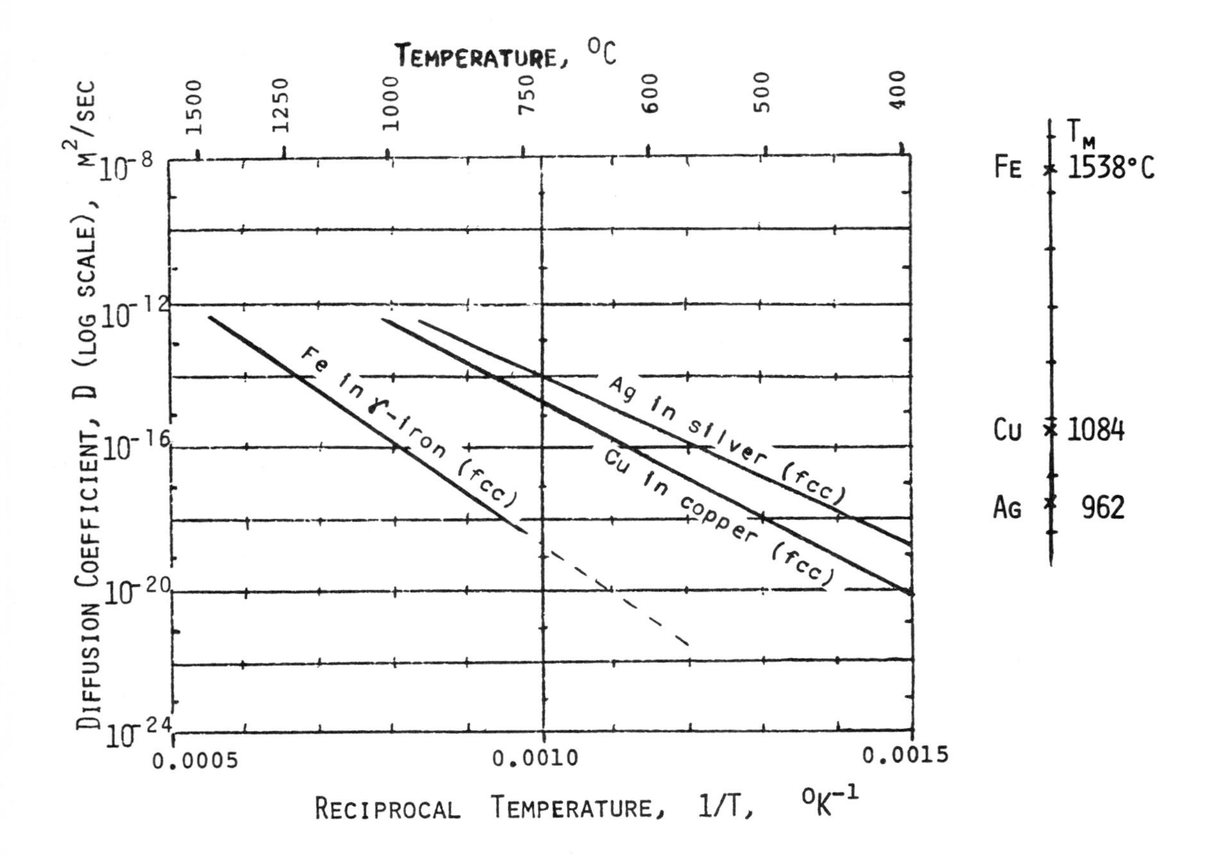

10. **Diffusivity *vs.* bond strength** (or melting temperature). The self diffusion of iron in fcc iron (γ) is shown in the lower curve of this figure. The comparable curves for copper and silver are also included. A comparison is useful. In each case, the host structure is fcc, so there is no difference in the atomic packing factor, as there was between bcc and fcc iron. In each case the diffusing atom is the same size as the solvent atoms, so interstitial diffusion is essentially absent. There is, however, a significant difference. The *bonding forces* between the adjacent atoms of the three materials are different. We have good evidence of this. The *melting temperatures* of Ag, Cu, and Fe are 960°C, 1084°C, and 1538°C, respectively. Silver is not as strongly bonded as iron or copper. Not surprisingly, we find that for any particular temperature, the diffusion coefficient for silver is above that for copper, and that for copper is greater than the diffusivity for high-melting iron.

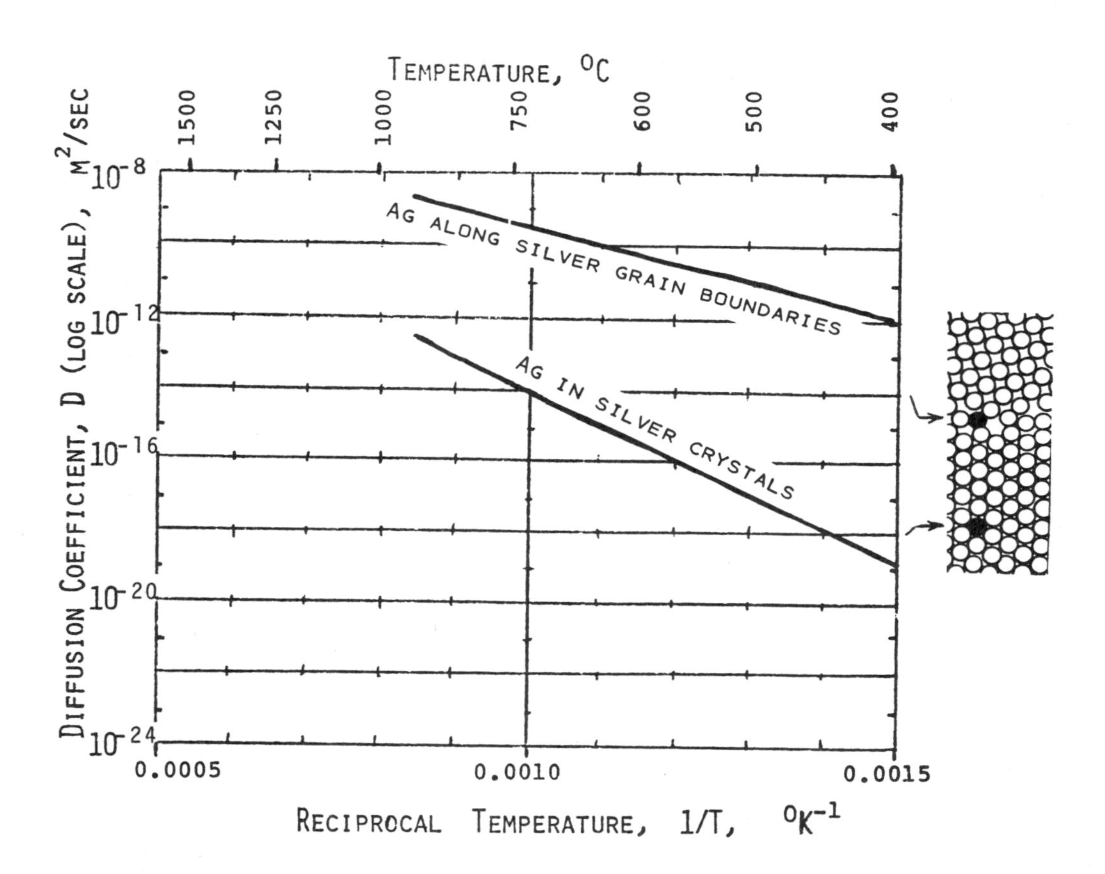

11. **Boundary diffusion.** Let us make one final comparison. Here we see the self diffusion coefficient for silver atoms *within* a silver *crystal* as the lower curve. The upper curve is for self-diffusivity of silver atoms *along the boundary* between two grains of silver. These two paths of movements are illustrated schematically at the right. The boundary provides a rapid diffusion path, because the zone of mismatch is inefficiently packed.

	FASTER DIFFUSION	SLOWER DIFFUSION
TEMPERATURE	HIGHER TEMP.	LOWER TEMP.
DIFFUSING ATOM	INTERSTITIAL ATOM (E.G., C IN FE)	SUBSTITUTIONAL ATOM (E.G., NI IN FE)
HOST STRUCTURE (PACKING)	LOWER A P F (E.G., BCC)	HIGHER A P F (E.G., FCC)
HOST STRUCTURE (BONDING)	WEAKER BONDS (E.G., LOW T_M)	STRONGER BONDS (E.G., HIGH T_M)
MICROSTRUCTURE	ALONG GRAIN BOUNDARIES	THROUGH CRYSTALS

12. **Summary of factors affecting diffusivity in metals.** Now we can summarize. When other things are equal:

1) atoms move more rapidly at *high temperatures* than at *low temperatures* ;
2) *smaller atoms* diffuse more rapidly than *large atoms*, because they can move interstitially;
3) atoms move less readily in *closely packed structures*;
4) they travel preferentially along *grain boundaries*;

and 5) they diffuse more readily in *weakly bonded* (low-melting) materials.

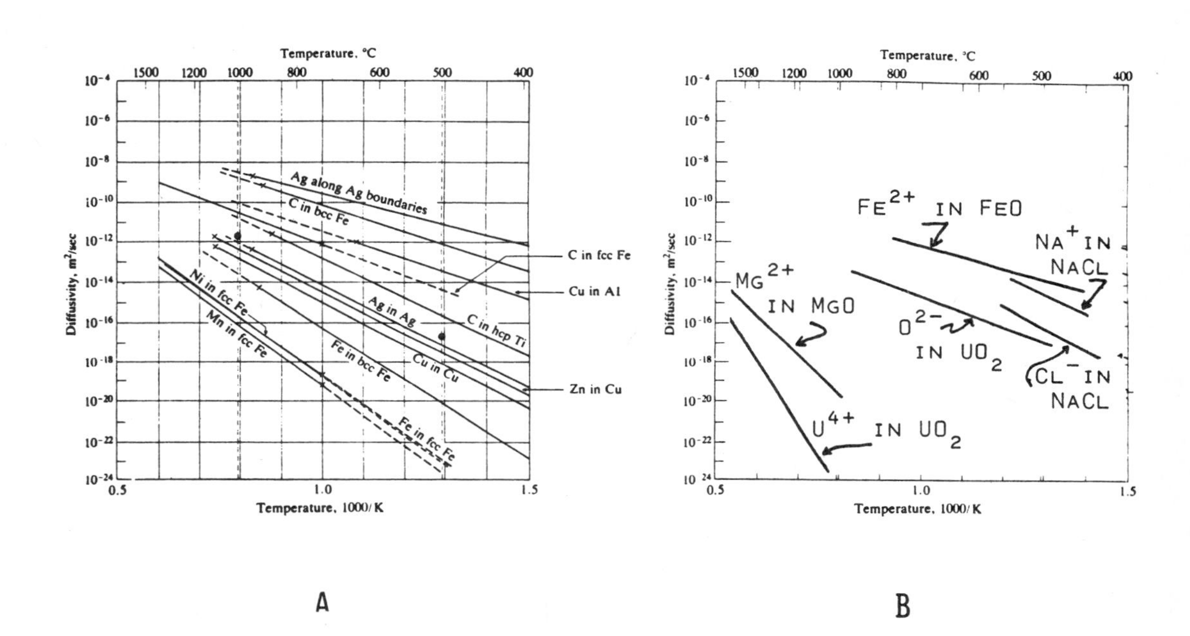

13. **Ionic diffusion.** This figure combines the previous diffusivity (diffusion coefficient) data for several examples of metallic diffusion (A) with some examples of ionic diffusion in ceramic oxides (B). Several comparisons among the latter can be explained.

1) We know that Na^+ ions in NaCl are only 60% as *large* as are Cl^- ions; therefore, $D_{Na^+} > D_{Cl^-}$.

2) Although NaCl and MgO have the same structure, the *double charges* of Mg^{2+} and O^{2-} produce a very strong, nearly immovable bond between the two components of MgO. Thus, $D_{Mg^{2+}} >> D_{Na^+}$.

3) There are many O^{2-} *vacancies* within UO_2; thus, even though the O^{2-} ion is larger than the U^{4+} ion, it can move more readily than the metal ion.

4) Finally, Fe^{2+} ions diffuse easily in ferrous oxide, because that oxide is really $Fe_{<1}O$ with many Fe^{2+} *vacancies.*

$$D = D_0 e^{-E/kT} \qquad (1)$$

$$\ln D = \ln D_0 - E/kT \qquad (2)$$

$$\log D = \log D_0 - E/2.3kT \qquad (3)$$

$$\log D = A - B/T \qquad (4)$$

14. **Arrhenius equations**. We should back up and look at the relationship between the diffusion coefficient (diffusivity), D, and temperature, T, more closely. Since the relationship between these two variables is not linear, we simply gave the equation, $\log D = A - B/T$ without explanation. We can be more precise. Therefore, at a later date, we will be able to use related principles when we talk about other processes within materials. These will include thermally related reactions such as *semiconduction*, *stress relaxation*, *grain growth*, and *corrosion rates*.

The more basic relationship for diffusion is shown in Eq. (1) above. There, D, is proportional to the exponential, $-E/kT$; and D_0 is the proportionality constant. Before we identify E and k, let us rearrange this equation to a logarithmic form (Eq. 2 or 3). Finally, the simplification to $\log D = A - B/T$ in Eq. (4) is achieved by equating the constant, $\log D_0$ to A, and the ratio, $E/(2.3)k$ to B. Of course, natural logarithms can also be used. However, when using plotted data, as we have in the preceding sketches, the $\log_{10}$ basis is more convenient.

So much for algebra. Now let's look at Eq. (1) again, which contained $-E/kT$, and proceed to the next page.

$$D = D_0 e^{-E/kT}$$

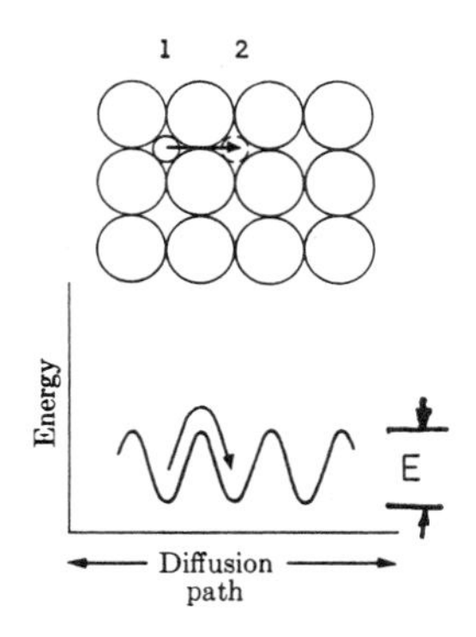

Constants for Diffusivity Calculations* ($\ln D = \ln D_0 - Q/RT = \ln D_0 - E/kT$)†

Solute	Solvent (host structure)	D_0, m²/sec	Q, cal/mole	E, J/atom
1. Carbon	fcc iron	0.2×10^{-4}	34,000	0.236×10^{-18}
2. Carbon	bcc iron	2.2×10^{-4}	29,300‡	0.204×10^{-18}
3. Iron	fcc iron	0.22×10^{-4}	64,000	0.445×10^{-18}
4. Iron	bcc iron	2.0×10^{-4}	57,500	0.400×10^{-18}
5. Nickel	fcc iron	0.77×10^{-4}	67,000	0.465×10^{-18}
6. Manganese	fcc iron	0.35×10^{-4}	67,500	0.469×10^{-18}
7. Zinc	Copper	0.34×10^{-4}	45,600	0.317×10^{-18}
8. Copper	Aluminum	0.15×10^{-4}	30,200	0.210×10^{-18}
9. Copper	Copper	0.2×10^{-4}	47,100	0.327×10^{-18}
10. Silver	Silver (crystal)	0.4×10^{-4}	44,100	0.306×10^{-18}
11. Silver	Silver (grain boundary)	0.14×10^{-4}	21,500	0.149×10^{-18}
12. Carbon	hcp titanium	5.1×10^{-4}	43,500	0.302×10^{-18}

* See J. Askill, *Tracer Diffusion Data for Metals, Alloys, and Simple Oxides*, New York: Plenum (1970), for a more complete listing of diffusion data.

† $R = 1.987$ cal/mole·K; $k = 13.8 \times 10^{-24}$ J/atom·K.

‡ Lower below 400°C.

A B

15. **Activiation energy.** It takes energy to move an atom from point 1 to point 2 among the surrounding atoms (A), since the adjacent atoms must be separated. This energy is called *activation energy*, *E*. Any specific atom has sufficient energy for movement only a small fraction of time. Boltzmann showed that this fraction is proportional to $e^{-E/kT}$. The value, k, is a constant (now called *Boltzmann's constant*), which relates the required energy to the absolute temperature, K. In SI units, its value is 13.8×10^{-24} J/K; (or in the eV energy units, 86.1×10^{-6} eV/K).

The diffusion coefficient is proportional to the numbers of atoms that have the above activation energy, with D_0 as the proportionality constant. Since the activation energy, *E*, is specific for each solute-host pair, we can list the activation energies. [See the last column of the table in (B).] Values of the pre-exponential constant, D_0, are also available. From these data, we can calculate the diffusivity at any selected temperature.

Based on data in the table, you calculate the value of *D* for copper diffusing in aluminum at 500°C. Turn to the next two pages to check your answer, as well as for several added comments.

A D OF CU IN AL AT 500°C (SOLUTION)

$$\ln D = \ln(0.15\times10^{-4})\ m^2/s - \frac{0.21 \times 10^{-18}\ J/\text{ATOM}}{(13.8 \times 10^{-24}\ J/°K)(773°K)}$$

$$= -11.1 - 19.7 = -30.8$$

$$D = \sim 4 \times 10^{-14}\ m^2/s$$

B D OF CU IN CU AT 500°C

$$\ln D = \ln(0.2\times10^{-4})\ m^2/s - \frac{0.33 \times 10^{-18}\ J/\text{ATOM}}{(13.8 \times 10^{-24}\ J/°K)(773°K)}$$

$$= -10.8 - 30.9 = -41.7$$

$$D = \sim 8 \times 10^{-19}\ m^2/s$$

15 (con't). Solution for Cu in Al diffusion calculation from p. 145. See comments on the opposite page.

C

$$\ln D = \ln D_0 - Q/RT$$
$$\log D = \log D_0 - Q/2.3RT$$
$$= \log D_0 - Q/4.575T \qquad \text{since } R = 1.987 \frac{\text{cal}}{\text{mole}\cdot{}^\circ\text{K}}$$

16. **Solution and comments** (pp. 145, 146). The answer for the value of the diffusivity of Cu in Al at 500°C is determined by the calculations in (A) on the opposite page. The diffusion coefficient for copper moving through copper (B) is also calculated from the data in the table on page 145.

Before we sign off on diffusivity (diffusion coefficients), let us cite one limitation for this type of calculation; then we will supply a final variant of the Arrhenius equation that is commonly encountered.

These equations (unless modified) apply only to *dilute solid solutions*. That is, the solute must be minor (usually , 1 a/o) for the data to hold. For example, a few copper atoms in aluminum will "see" only aluminim atoms. If there are a lot of copper atoms dissolved in the aluminum, the moving atoms will encounter a signification number of other copper atoms en route. Available data show that the copper atoms move less readily among copper atoms than among aluminum atoms. We should, therefore, expect that the diffusion coefficient will change with major increases in solute concentrations.

Above. Finally, a variant of the Arrhenius equation using Q/R is shown in (C). Commonly, the chemist speaks of the activation energy, Q, required for a mole (0.6 x 10^{24}) of material. Calories are usually used when this is done. Thus, Q is expressed as calories per mole. This requires a change from k to a different constant, R, to relate the activation energy to absolute temperature. The value of R is 1.987 cal/mole·K. You may recognize this as the familiar "gas constant" that is used in the gas-law equation, $PV = nRT$. Values for Q are given in the table on page 145.

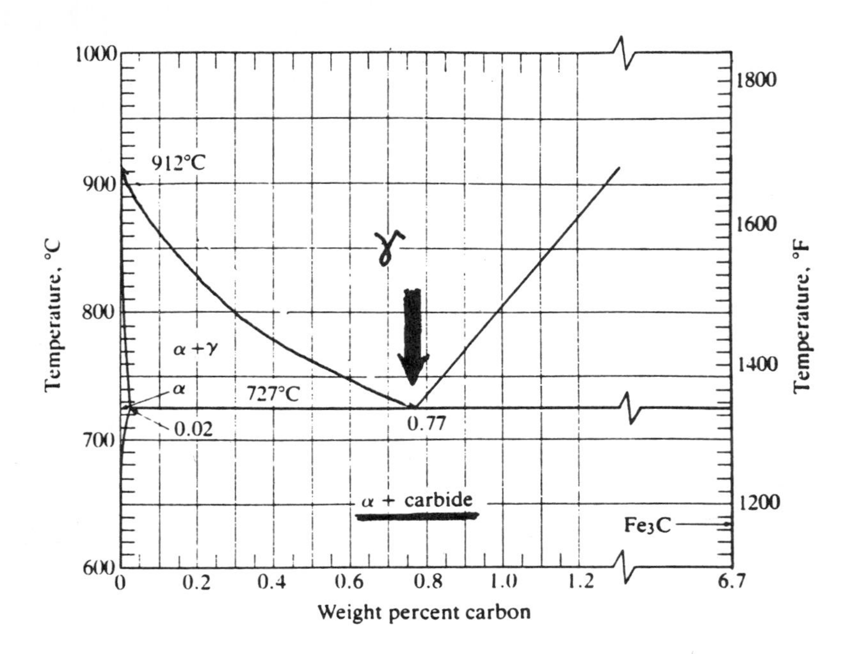

1000
900
800
700
600
912°C
γ
α + γ
α
727°C
0.02
0.77
α + carbide
Fe₃C
1800
1600
1400
1200
Temperature, °C
Temperature, °F
0
0.2
0.4
0.6
0.8
1.0
1.2
6.7
Weight percent carbon

IV

AUSTENITE

DECOMPOSITION

$\gamma \longrightarrow (\alpha + \text{carbide})$

1. **The eutectoid reaction.** A plain-carbon steel of eutectoid composition (nearly 0.8% carbon) is completely *austenite* if it is held above 727°C. All of the carbon is dissolved in this fcc iron (γ). The carbon atoms occupy the centers of the face-centered cubes (or the midpoints along the edges, which are crystallographically identical). However, since 0.77 w/o is only 3.5 a/o, there is one carbon atom for every seven fcc unit cells.

   Although only 3.5 a/o carbon is present, if the austenite is cooled to just below 727°C, it rejects almost all of the carbon, which forms *iron carbide* (Fe_3C). The remaining iron changes to ferrite, the bcc polymorph (α). The reaction at 727°C can be represented by the equation:

$$\gamma\ (0.775\ \text{C}) \xrightarrow{727^\circ\text{C}} \alpha\ (0.02\%\ \text{C}) + Fe_3C\ (6.7\%\ \text{C}).$$

   The ferrite, or α-phase, is bcc iron containing only a minute amount of carbon (0.02 w/o), which means only one carbon atom for more than 1100 iron atoms. The Fe_3C, on the other hand, contains one carbon atom to every three iron atoms. It has a complex crystal structure, and is both hard and brittle.

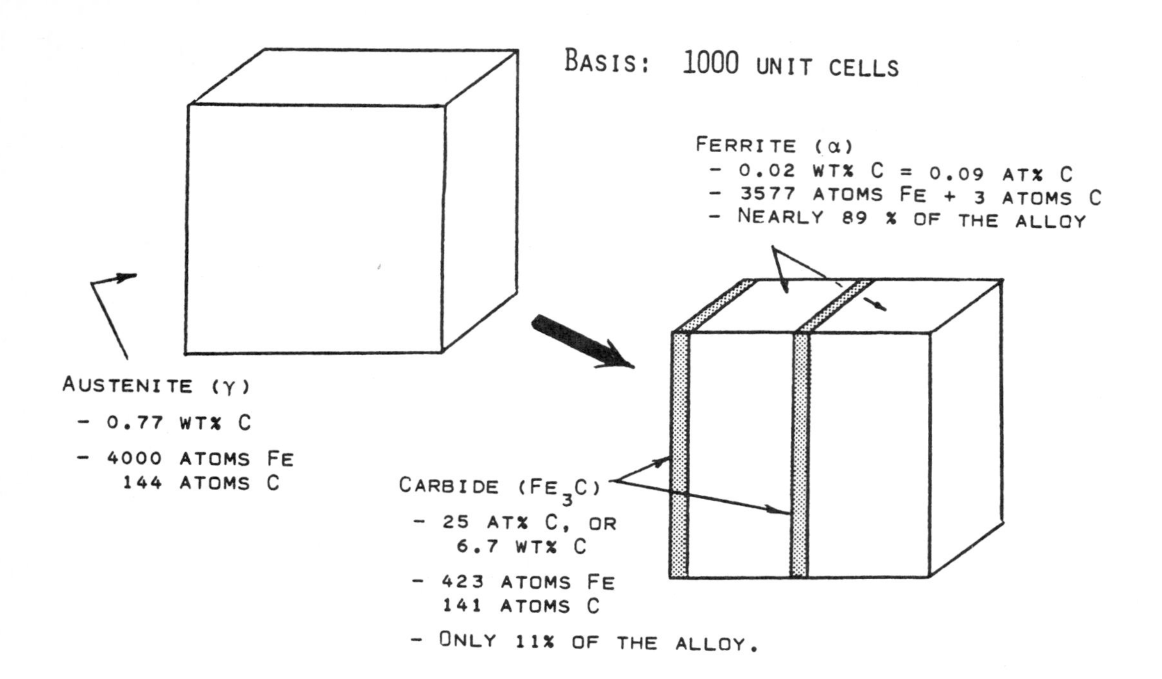

2. **Composition changes.** The composition changes that occur in the eutectoid reaction are quite dramatic. To understand this, visualize one thousand unit cells in a block of 10 x 10 x 10 unit cells of austenite (γ). This block contains 4000 iron atoms and 144 carbon atoms (0.77 w/o, or 3.5 a/o) .

When the austenite transforms, 141 of these 144 carbon atoms go into the carbide phase, leaving only three of them in the bcc iron phase. The relative amounts of the two phases, α and Fe_3C, can be calculated from the phase diagram, using the lever rule.

$$\text{Wt\% } Fe_3C = 100(0.77 - 0.02)/(6.7 - 0.02) = 11.2\%.$$

So what happens is that 98 percent of the carbon atoms (141 out of 144) go into a small fraction (11.2%) of the total product. For this to happen, the carbon atoms must move, migrate, or diffuse, from one part of the metal to another. (Since diffusion takes time, it is possible to suppress the transformation by quenching.)

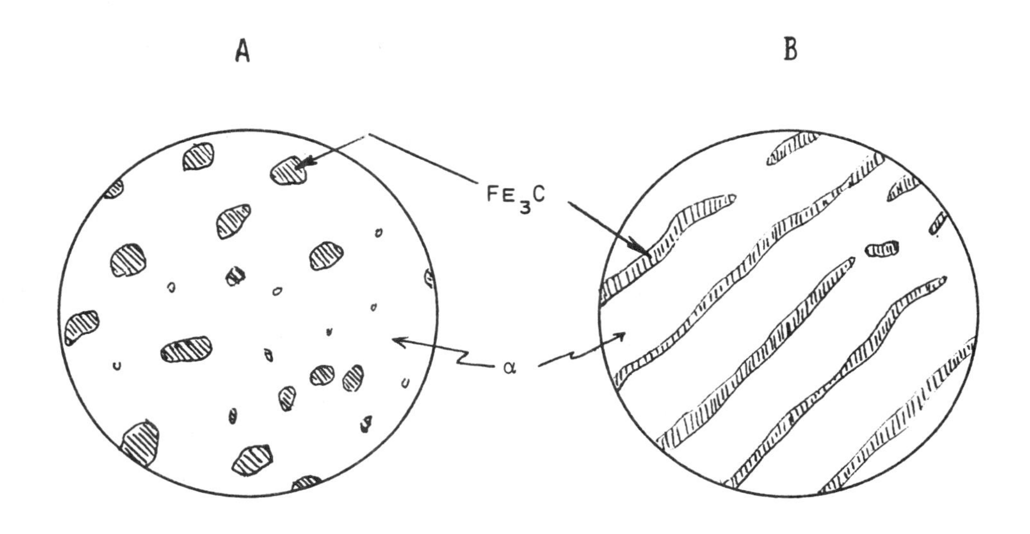

SPHEROIDITE

-PARTICLES OF CARBIDE IN A FERRITE (α) MATRIX.
-APPROACHING AN EQUILIBRIUM STRUCTURE.

PEARLITE

-LAMELLAE OF FERRITE (α) AND CARBIDE.
-FORMED BY EUTECTOID DECOMPOSITION OF AUSTENITE (γ).

3. **Phase distributions.** If we hold our eutectoid steel at a temperature just below 727°C for a long time so that the metal is equilibrated, both α and carbide will be present. They appear as spheroidal particles of carbide distributed in a matrix of ferrite (A). We call the microstructure *spheroidite.* It is a stable, low-energy structure. (You are aware that spheres have the smallest surface area per unit volume of any 3-dimensional shape; so spheroidized particles minimize the boundary energy between the two phases.)

However, if we cool our eutectoid steel slowly through the eutectoid temperature, or allow it to transform isothermally at, say, 650°C, we find that the (α + carbide) structure is like (B) above. The carbide is not present as spheres; rather it has a characteristic lamellar arrangement of ferrite and carbide plates, which we call *pearlite.* This microstructure is found in the majority of steels that are not quenched.

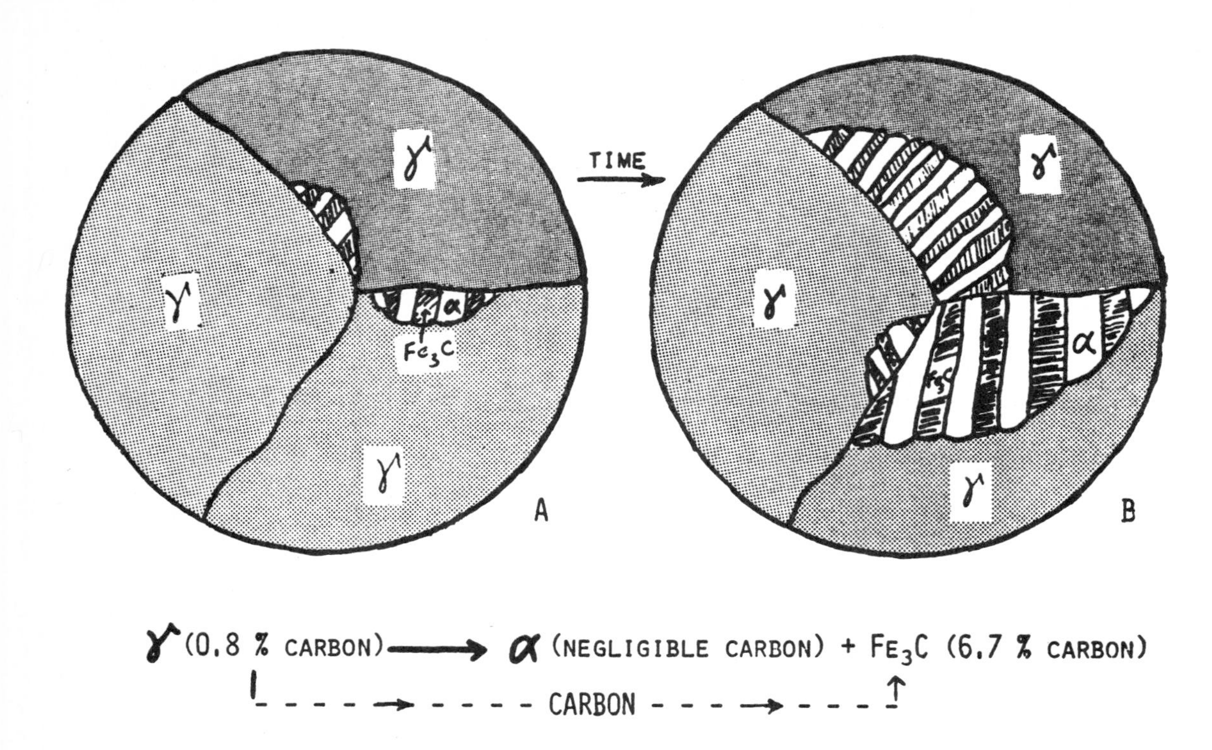

4. **Growth of pearlite.** The two phases of pearlite form alternating layers because the γ-decomposition can start at relatively few nuclei along grain boundaries and then grow into the austenite grain. In contrast, to have a dispersion of spheroidal carbide particles in an α matrix, it would be necessary for each carbide particle to be nucleated separately. The formation of the plates is consistent with the theory of nucleation and growth for transformation products.

   The plates of α and carbide have a definite crystallographic relationship to the former austenite grains, and to each other. This encourages the growth of parallel plates rather than some other shape, such as rods. This favored growth arises because this mutual relationship produces a minimum of "mismatch" along the interface between the adjacent phases.

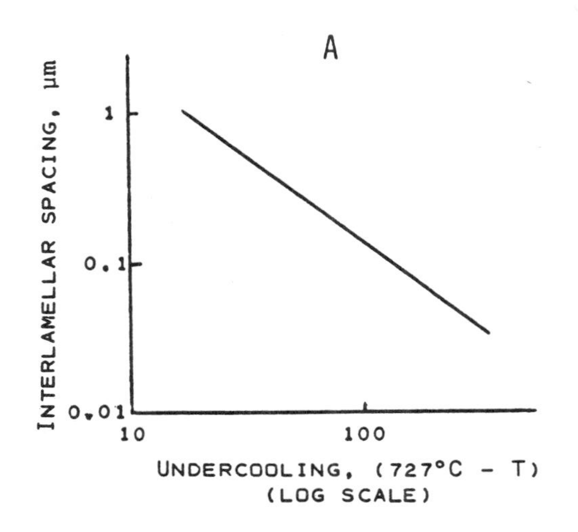

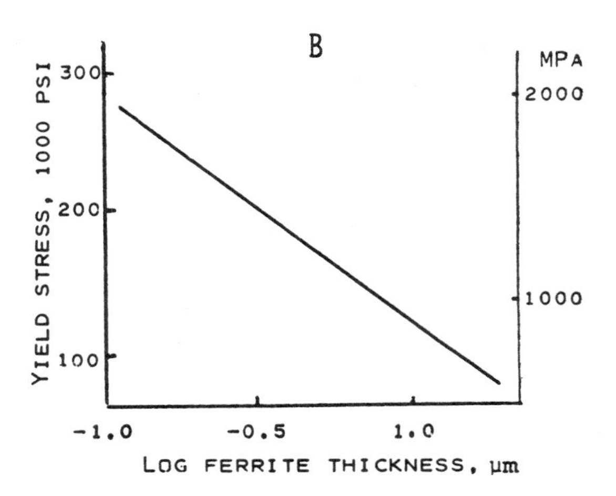

5. **Pearlite spacing.** Ferrite (α) and carbide are present in pearlite in roughly an 8/1 ratio. The only other parameter is the thickness of the lamellae. This is controlled by the temperature at which the pearlite forms.

If pearlite is formed at a *constant temperature* , the higher the temperature of formation, the coarser the pearlite. In fact, theory predicts that the interlamellar spacing of pearlite formed at a temperature, *T*, should vary inversely with the amount of undercooling below the eutectoid temperature (= 727 - *T*)°C. This is shown in (A).

Similarly, if the pearlite is formed during a *continuous cooling* process, the slower the rate of cooling , the higher will be the average temperature of transformation, and the coarser will be the pearlite. For example, steel may be cooled in a furnace from its austenitizing temperature. This slowly cooled steel transforms early in the cooling process and produces coarse pearlite, in contrast to steel removed from the furnace and cooled in air.

The pearlite spacing is important because it has a marked effect on the hardness and strength of the steel (B). Finer pearlite is harder and stronger. Conversely, a steel with a coarser microstructure is more readily machinable.

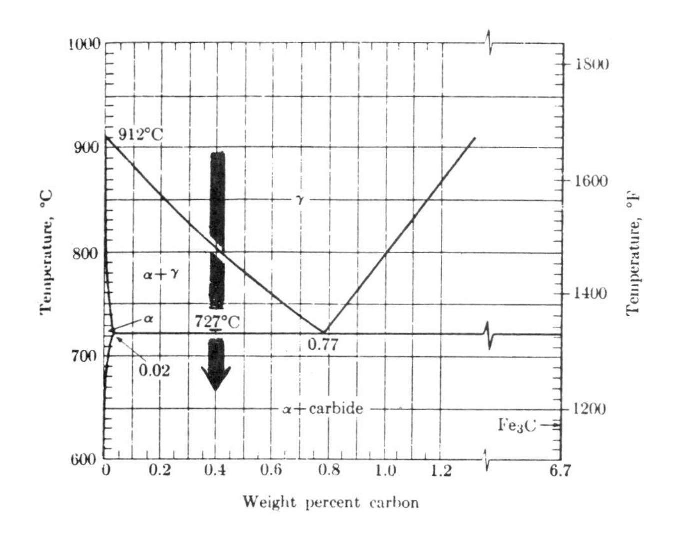

A

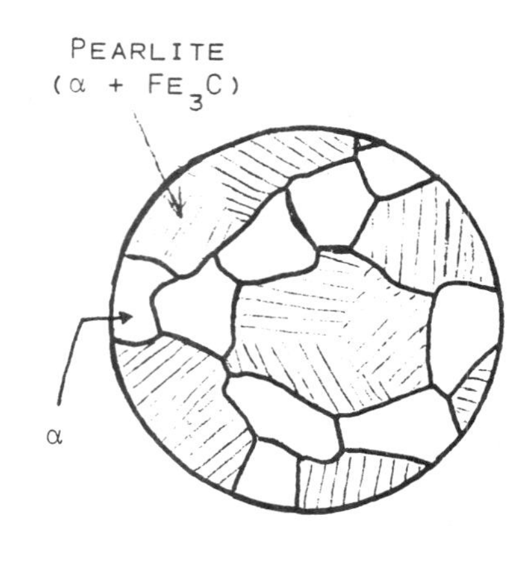

B

6. **Hypoeutectoid steel.** So far we have discussed plain-carbon steels containing about 0.8% carbon,-- eutectoid steels. The situation is comparable with either higher *or* lower carbon contents, except that another phase always appears *before* the pearlite appears. Let us examine a *hypoeutectoid* steel, i.e., a steel with less than eutectoid carbon content.

If the 0.4% carbon steel of (A) is cooled slowly from an austenitizing temperature of 850°C, some α-phase (ferrite) starts to form at about 800°C, and the amount of α increases as cooling proceeds. At the eutectoid temperature, the steel is roughly half α and half γ. However, it is important to realize that the γ now contains 0.8% carbon! As the low-carbon α is formed, the carbon remaining in the γ increased from the original 0.4% to the eutectoid composition (0.8% C) at the eutectoid temeprature (727°C).

With further cooling below the eutectoid temperature, this γ of eutectoid composition transforms to (α + Fe_3C) in the form of pearlite, just as the 0.8%-carbon steel did. The final microstructure looks something like that shown in (B). It contains α that formed first (*proeutectoid*). It is intermixed with, and to some extent surrounds, the regions of pearlite that subsequently formed below the eutectoid temperature.

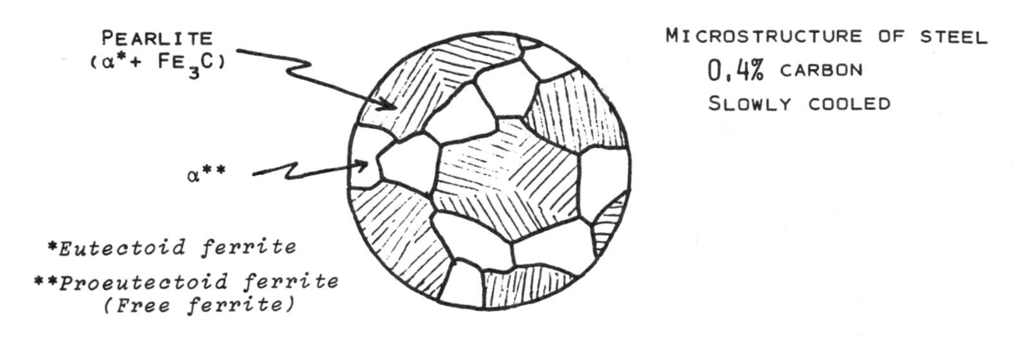

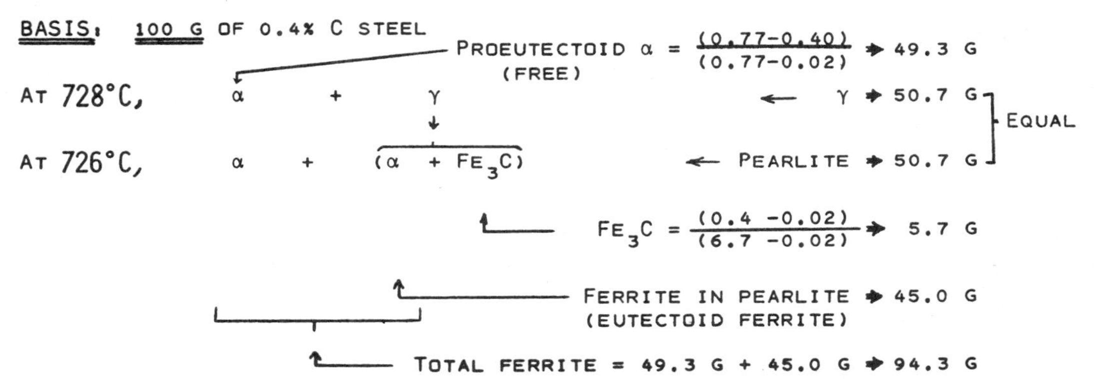

7. **Proeutectoid ferrite and total ferrite.** The Fe-C phase diagram says only that the phases in equilibrium below 727°C in our 0.4%-carbon steel are α and Fe_3C. It does not distinguish between the α, which formed before the eutectoid reaction, and the α, which is present in the pearlite. We shall call the former *free ferrite,* or *proeutectoid ferrite,* and the latter *eutectoid ferrite.*

It is a simple matter to calculate the amounts of α and Fe_3C from the lever rule. The total ferrite (α) at 726°C is 94.3%, and Fe_3C is 5.7%.

However, it can be important for us to know how the α-phase is distributed between the pearlite and the proeutectoid ferrite. We can determine this by applying the lever rule to the ($\alpha + \gamma$) region at 728°C, slightly above the eutectoid temperature. We know that the percent austenite at 728°C will be the same as the percent pearlite in the final structure. In this way, we get the amount of *pearlite* to be 50.7%, and the *free ferrite* to be 49.3%.

Similar arguments can be applied to the discussion of *hypereutectoid* steels, i.e., those with more than 0.8% carbon, the only real difference being that the proeutectoid phase would be *cementite* (Fe_3C) instead of ferrite (α) treated above.

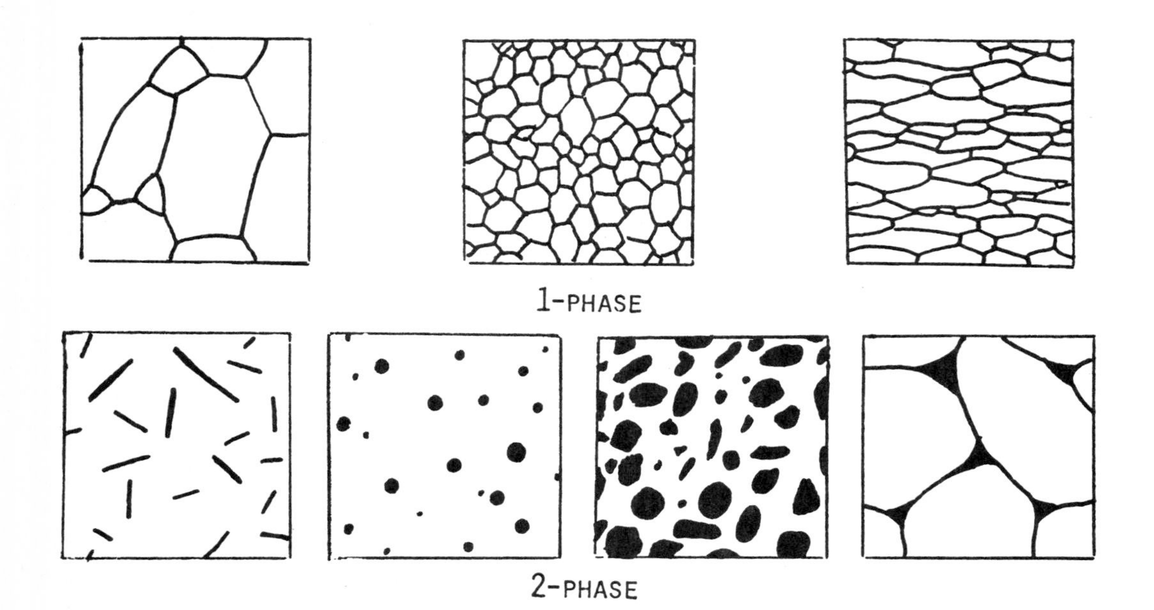
1-PHASE
2-PHASE

V

MICROSTRUCTURES

of

STEELS

1. **Microstructure of solids** (opposite). By the term, *microstructure* , we refer to the geometry of phases and grains within a solid material. Typically, but not always, magnification is required for us to observe microstructures visually.

   These sketches show a variety of microstructures. The first three reveal variations of *size* and *shape* of grains in a *single-phase* material. The remaining sketches are various microstructural geometries of materials with *two phases*. Variations are found in the relative *amount*, the *shape*, and the *distribution* of the two phases.

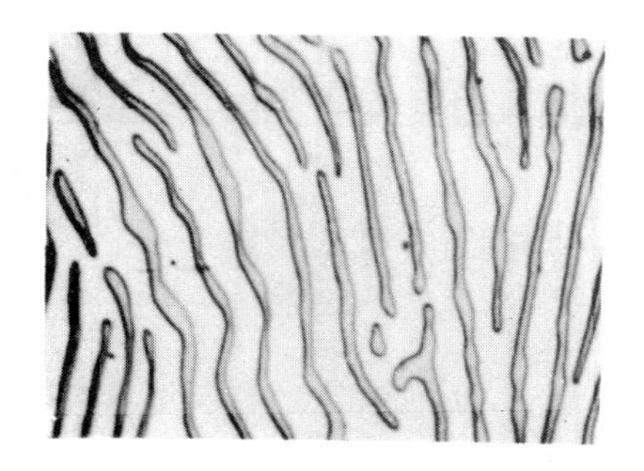

PEARLITE x2500

BAINITE x11,000

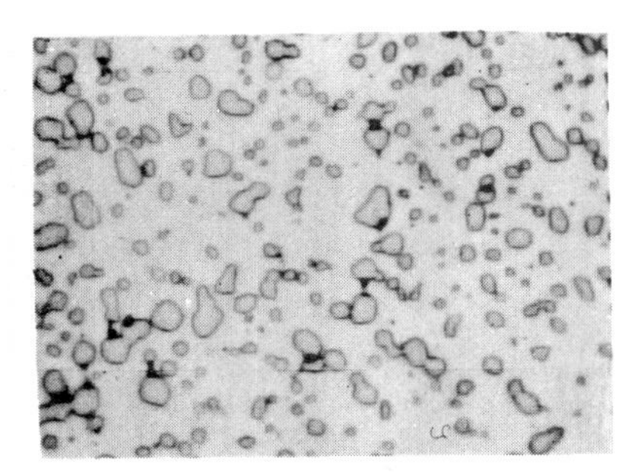

SPHEROIDITE x2500

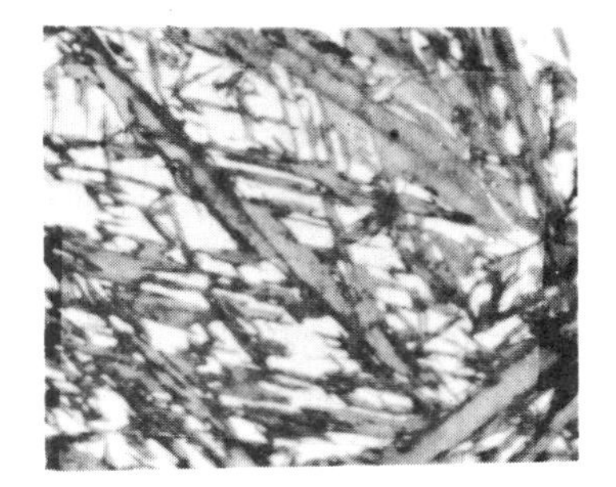

MARTENSITE x1000

2. **Examples of microstructures of steel.** The four photomicrographs on this page all have the same composition (99.2Fe-0.8C). Their microstructures, however, are obviously different. This is the result of different heat treatments. An important consequence is that the properties of the resulting materials are markedly different.

   The first microstructure has the given name of *pearlite* and contains alternate layers of ferrite (bcc Fe) and carbide (Fe_3C). It has intermediate strength and relatively low ductility. (U.S.Steel.)

   The second microstructure is called *bainite* with very fine carbide particles in a ferrite matrix. It is relatively strong and also tough, a combination that has desirable characteristics for engineering design. (General Motors.)

   The third photomicrograph also has a dispersion of carbide particles in a ferrite matrix; however, it is coarser, and therefore softer than bainite. We call it *spheroidite* because the carbide particles are globular. (U.S. Steel.)

   The final microstructure with this same composition has no ferrite nor carbide, but contains an extremely hard phase called *martensite* (gray), and some residual *austenite* (white). Martensite is extremely brittle. The brittleness excludes the use of this steel in most applications, even though its high hardness is attractive. (U.S. Steel.)

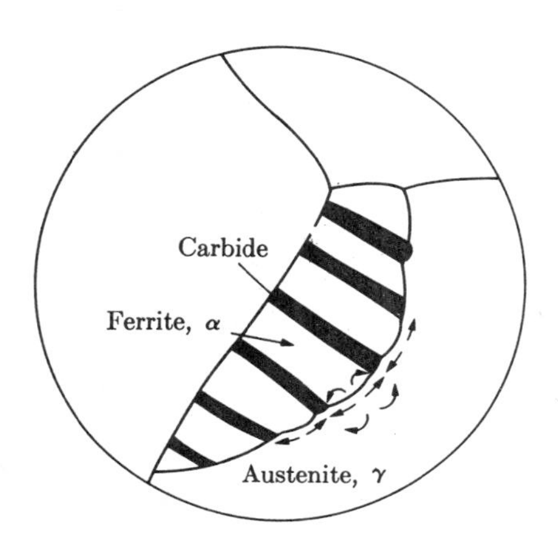

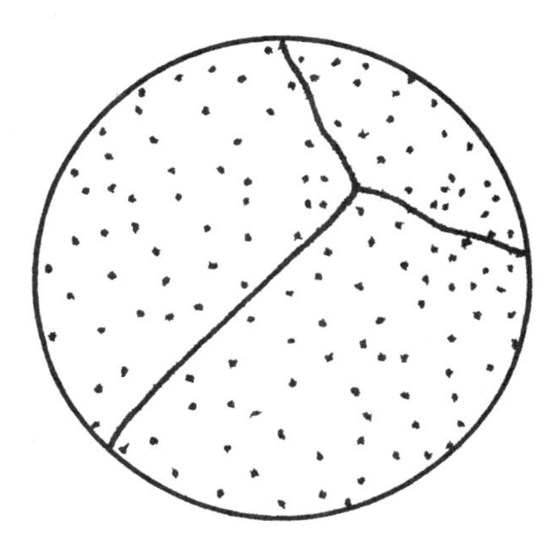

A

B

3. **Pearlite and bainite formation.** Pearlite and bainite are formed by *austenite decomposition* . Specifically, the iron-carbon alloy is heated into the austenite (γ) region of the phase diagram. On cooling,

$$\gamma \longrightarrow \alpha + \text{carbide.}$$

The products are ferrite (α) and carbide (Fe_3C).

Whether pearlite (layered carbides) forms, or whether bainite (particulate carbides) forms, is a consequence of how we permit the above reaction to occur. For *pearlite* to form, we must provide conditions so that the ferrite and carbide layers grow simultaneously from the boundary toward the interior of the former austenite grains (A). For *bainite* to form, the pre-existing austenite (γ) must be cooled to a lower temperature before the reaction starts. Since diffusion is slow, carbide nucleation occurs at numerous imperfection sites within the former austenite grains (B).

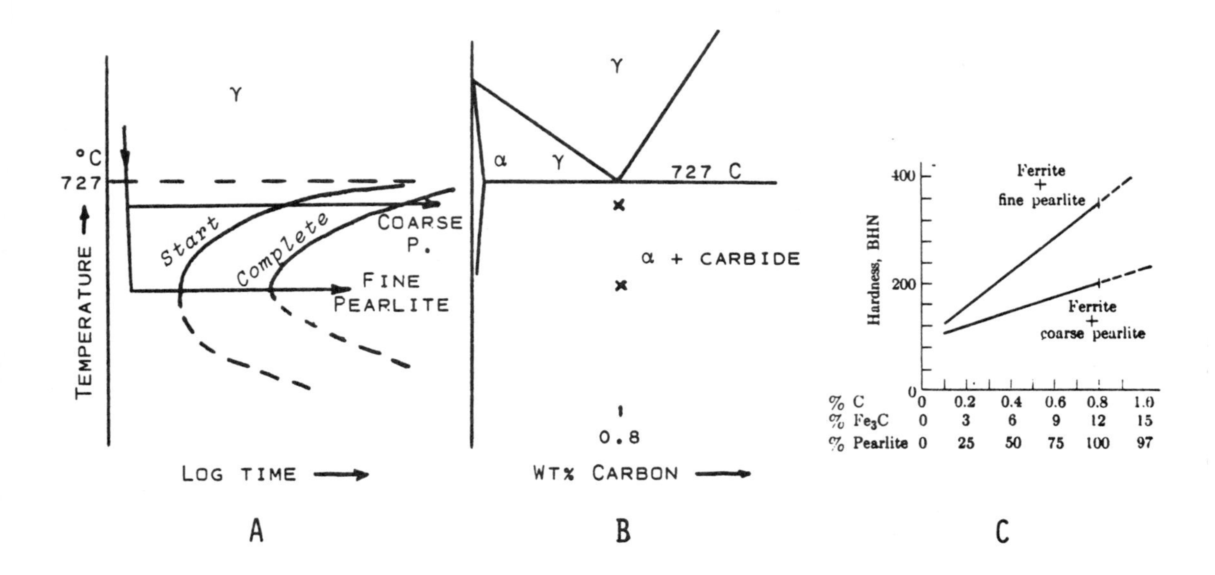

4. **Coarse and fine pearlite.** The effect of temperature on pearlite growth is shown in (A). At temperatures just under the eutectoid [just under 727°C of (B)], more time is required. The resulting layers of pearlite are relatively thick. At somewhat lower temperatures (A), pearlite forms in shorter times; the resulting layers of pearlite are thin. This *fine* pearlite is stronger, harder (C), and less ductile than the coarse pearlite.

Pearlite also forms when steel is cooled slowly during the ($\gamma \longrightarrow \alpha$ + carbide) reaction. Delayed cooling produces very coarse pearlite; normal cooling rates produce finer pearlite.

X11,000

5. **Bainite** (optional). The effect of temperature on bainite formation is shown in the above figure. At higher temperatures, a 1080 steel (0.8% C) forms bainite more slowly than it forms pearlite; therefore, we never have bainite form at temperatures above 500 or 600°C in this steel. At lower temperatures, however, bainite forms in less time than does pearlite (A). Therefore, we must cool the γ to temperatures below ~500°C in order to form bainite.

   Bainite formed at 300°C has many more (and much smaller) carbide particles than bainite formed at 500°C, because the ($\gamma \longrightarrow \alpha$ + carbide) reaction is nucleated at many more points throughout the original austenite grains. The low-temperature bainite is thus harder than the high-temperature bainite (B).

   The composite ($\gamma \longrightarrow \alpha$ + carbide) curve for the two products is shown in (C). The faster product wins; therefore, we see pearlite formation above the "knee" of the isothermal transformation curve, and bainite formation below the "knee" of the curve.

   The commercial process of bainite formation in the above manner is called *austempering*.

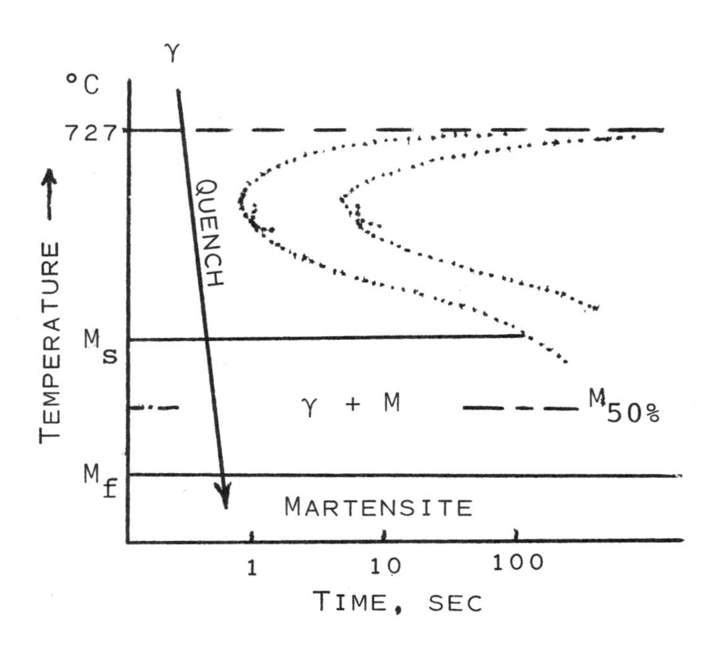

6. **Martensite formation.** *Martensite* forms in a 99.2Fe-0.8C steel by avoiding the ($\gamma \longrightarrow \alpha$ + carbide) reaction through a rapid quench (Q).

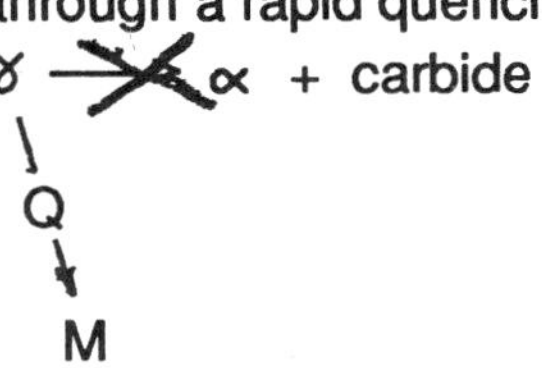

In the above steel, the quench from 750°C to the low temperature must occur in only a few seconds. This means, of course, that martensite can not be formed in the center of large parts made from this eutectoid steel, simply because we cannot extract the heat in the available time.

In the above 0.8% carbon steel, martensite starts to form (M_s) at ~300°C, if quenched rapidly. The temperature must be lowered to -40°C for the final, completed martensite formation (M_f).

γ ----X----> α + CARBIDE

Q ↘ M — TEMPER ↗

X11,000

7. **Tempered martensite.** Martensite is not a stable phase and does not provide a stable microstructure. In fact, it does not even appear on the phase diagram. Rather, martensite decomposes on heating to form (α + carbide). Thus, when the martensite is formed by quenching (Q), we can write

γ —X→ α + carbide

Q ↘ M ↗ T

We must heat the steel to ~200°C for the (M ——> α + carbide) leg of the reaction to proceed. We call this *tempering* (T), and we call the resulting (α + carbide) microstructure, *tempered martensite* . This microstructure possesses a very fine dispersion of carbide particles in a ferrite matrix. It is similar to the microstructure of bainite; thus, like bainite, it can be both hard and tough. Although similar to bainite, tempered martensite is formed through different process steps. Therefore, it is advantageous to retain the two distinct names,-- *bainite* and *tempered martensite. (Electron Microstructures of Steel,* A.S.T.M. and General Motors.)

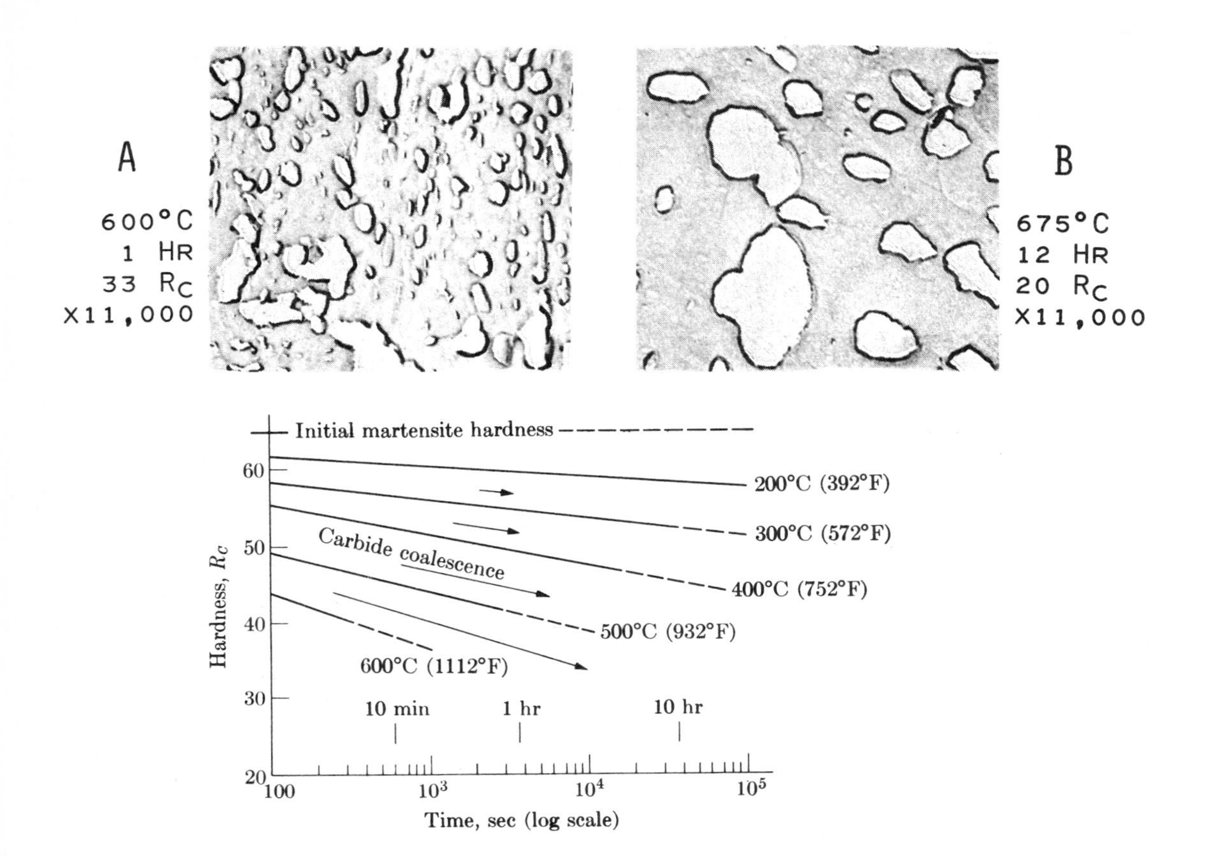

8. **Carbide coarsening.** The microstructure of any material coarsens if given an opportunity. Thus, the very fine carbide particles of tempered martensite (A) become larger, and fewer (B) if the steel is heated to an elevated temperature (but still below the 727°C eutectoid temperature). The distances between the particles are increased. Because the matrix is a soft ferrite, this microstructural change softens the steel. (*Electron Microstructure of Steel,* A.S.T.M. and General Motors.)

   At low temperatures, the rate of coarsening is very slow, usually undetectable. At higher temperatures, less time is required. As a result, the steels soften more rapidly as the temperature is increased. This is shown in (C).

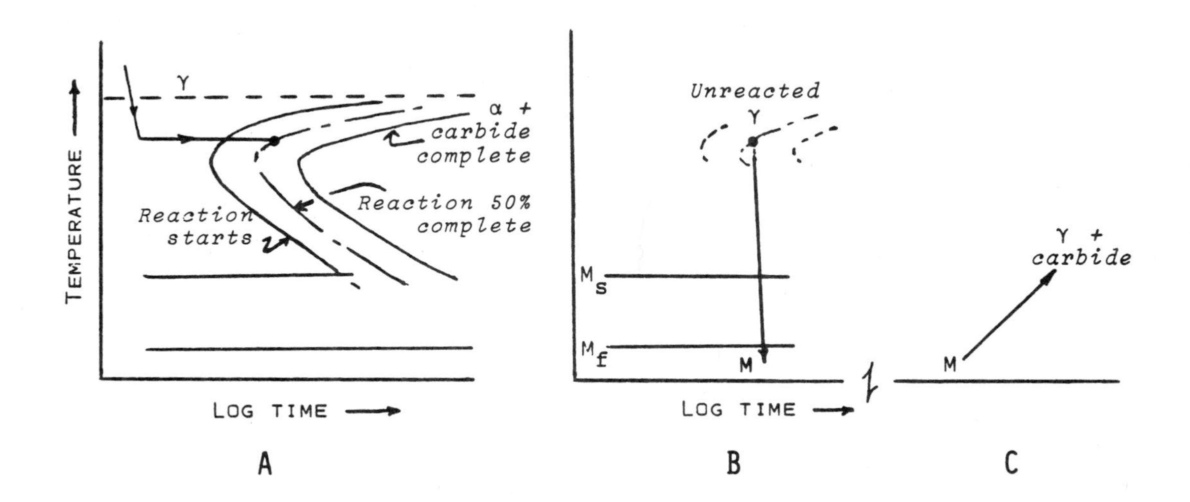

9. **Mixed microstructures.** With appropriate heat treatments, we may obtain mixtures of some of the previously described microstructures. For example, permit the (γ ——> α + carbide) reaction to transform half of the austenite to pearlite (A); then quench the steel so that the remaining γ goes to martensite (B). The result is a mixture of pearlite and martensite.

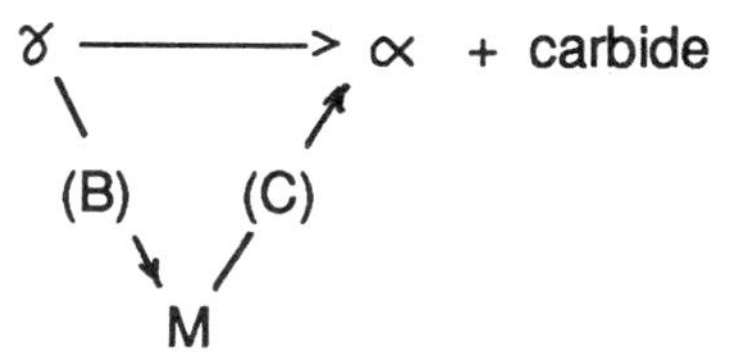

Of course, the martensite can now be tempered (C) to give a final microstructure of 50% pearlite and 50% tempered martensite.

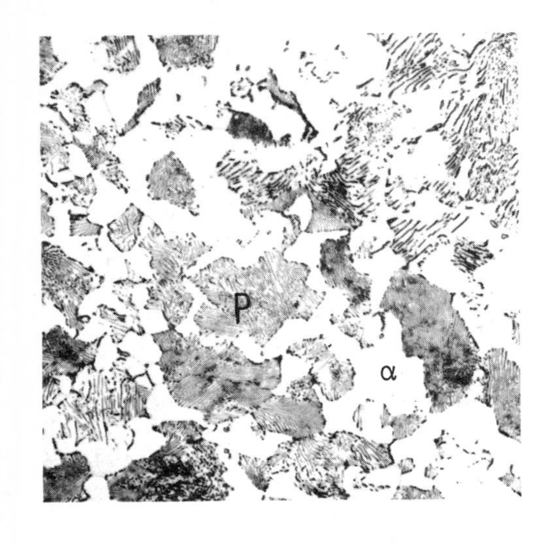

A

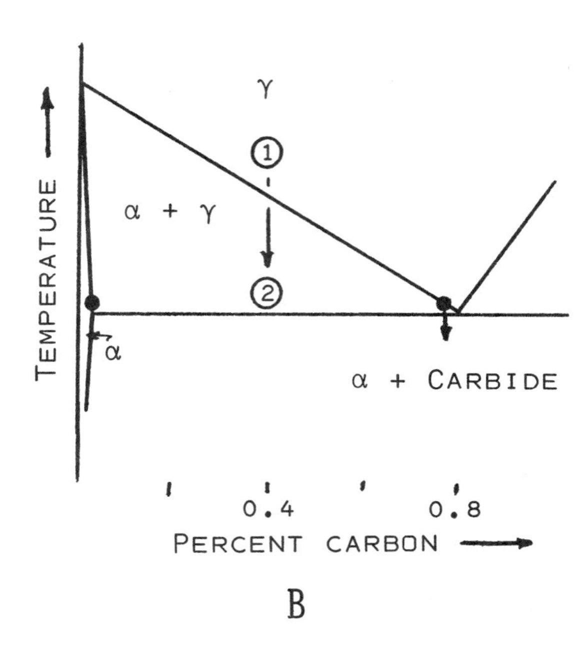

B

10. **(Non)eutectoid steels.** Mixtures of microstructures are also possible via compositional control. A 1040 steels (0.4% C) has only half of the carbon content of 1080 (eutectoid) steels. If 1040 steels are cooled slowly, a mixture of ferrite (∝) and pearlite (P) is developed in (A). (Magnification: X500.)

We may undertand this better if we (i) take a 1040 steel and form only austenite (γ), ① in (B); (ii) cool it to ② in (B), and hold it at that temperature until an equilibrated 50-50 ferrite-austenite mixture develops. Note that the γ at ② has 0.8% C; therefore, that half of the steel will behave in the same manner as a 1080 steel would when it is transformed to pearlite, bainite, martensite, or tempered martensite. The ferrite half of the steel simply gets colder during any subsequent heat treatment below 700°C. (U.S. Steel.)

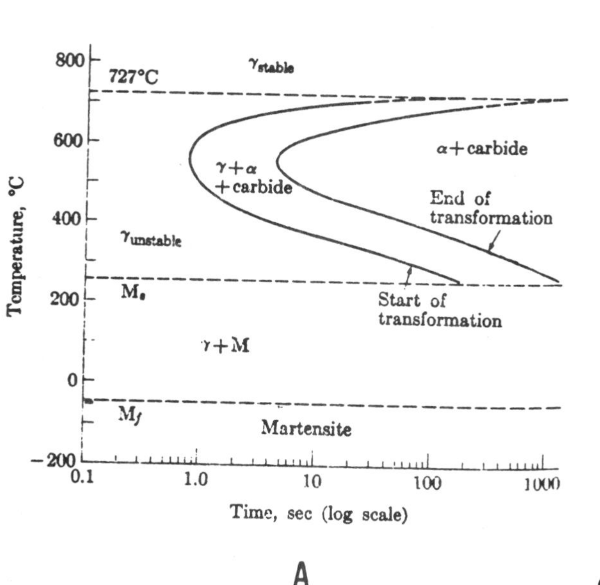

A

(1080)

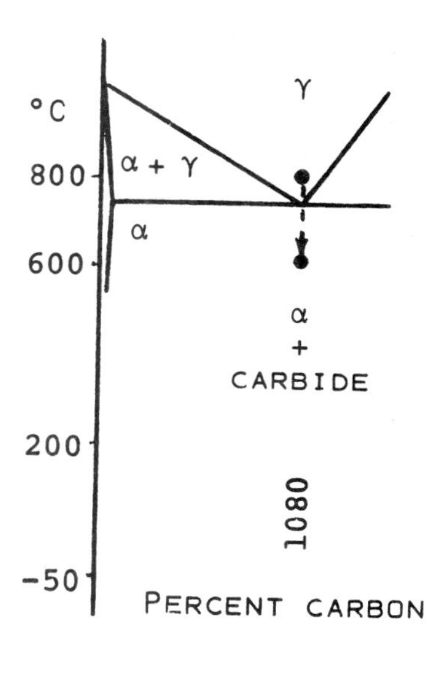

B

11. **Problem No. 1.** Now let us consider a series of heat treatments for a 1080 steel. You cite the microstructures after each sequential step, based on the data in the above sketch.

   a) Heat to 800°C until equilibrated.

   b) Quench to 600°C; hold 2 seconds.

   c) Quench to -50°C.

   d) Reheat to 200°C.

After you jot down your answers, turn to the next sketch and check yourself.

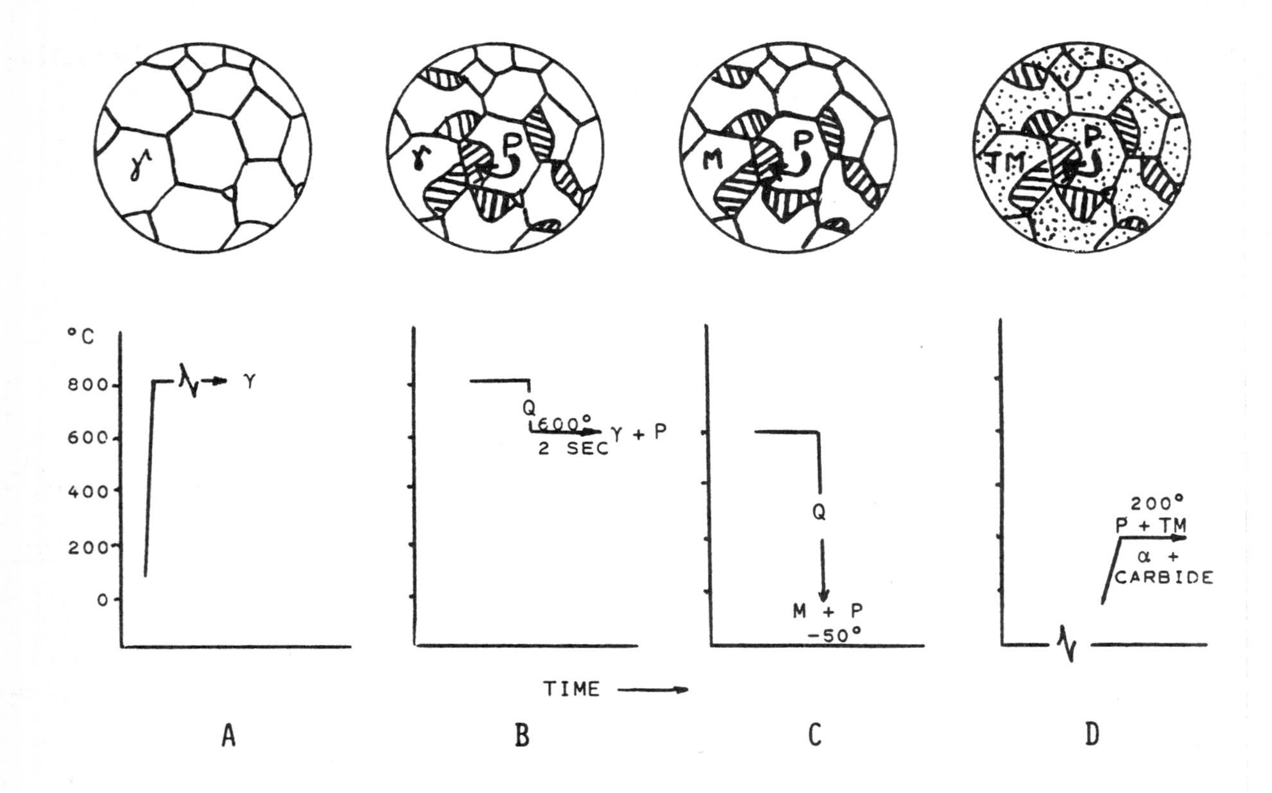

11 (con't). **Problem No. 1** (solution). The 1080 steel from the previous sketch equilibrates to a single-phase, γ, microstructure at 800°C (A).

A quench to 600°C, and a hold for 2 seconds permits the (γ —> α + carbide) reaction to start. According to the preceding sketch, however, there is not time for completion. Thus, we have a mixture of pearlite and untransformed austenite (B).

With a second quench to -50°C, the austenite of (B) forms martensite. The pearlite of the (γ —> α + carbide) reaction was already complete, so it does not change any further (C).

The reheating to 200°C does not affect the pearlite. However, the martensite proceeds in its reaction to produce tempered martensite (D). Neither the pearlite nor the martensite revert to austenite at 200°C. As indicated by the phase diagram, γ develops only above the eutectoid temperature.

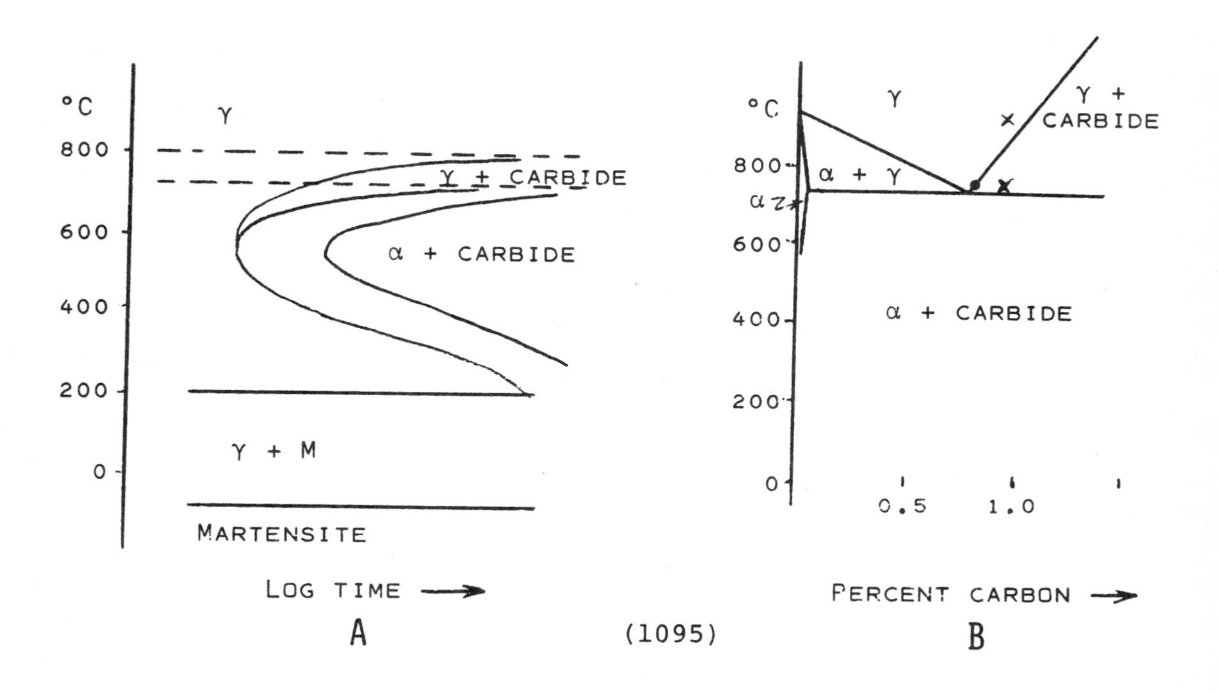

12. **Problems No. 2 and No. 3.** How about two more situations, each involving a 1095 steel, i.e., 0.95% C? This steel could be used in a razor blade. Use the above sketch and the sketch on p. 167.

2-a) Heat the 1095 steel to 900°C until equilibrated as austenite.

2-b) Quench to 20°C.

2-c) Reheat to 300°C.

Also, 3-a) Heat another sample of the same steel to 735°C until equilibrated.

3-b) Quench to 450°C; hold 10 seconds.

3-c) Quench to 20°C.

3-d) Reheat to 400°C.

Record your answers, then check yourself with the sketches on the next page.

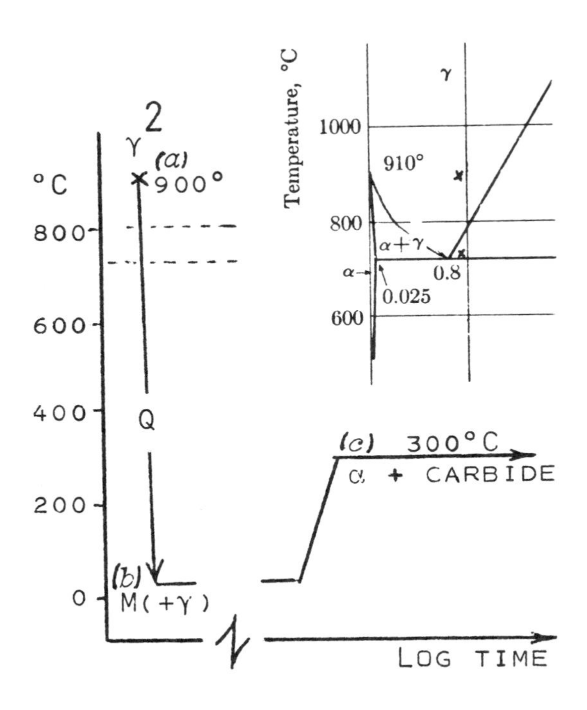

12 (con't) **Problem 2** (solution). According to the Fe-Fe_3C phase diagram, a 0.95% C steel is wholly γ at 900°C. According to the previous sketch (p.169), the quench to 20°C partially transforms the γ ——> M. On tempering it at 400°C, the martensite proceeds to (α + carbide) to produce tempered martensite. [Added note: With martensite removal, the previously retained austenite can either transform directly to (α + carbide), or to martensite and then to tempered martensite (α + carbide).]

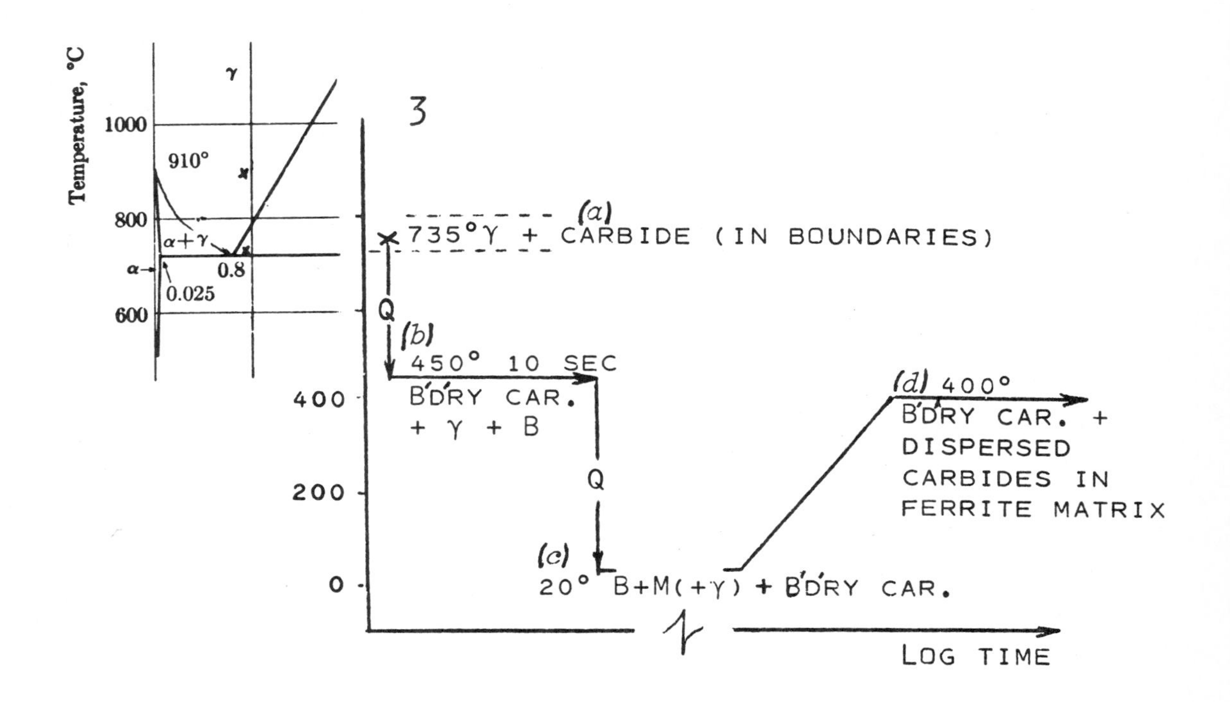

12 (con't) **Problem 3** (solution). At 735°C, and according to the phase diagram, a 1095 steel contains about 3% carbide. [Typically, this forms a boundary network around the austenite grains.]

The majority of the steel is now γ, which contains 0.8% C (and not 0.95% C). [If you are uncertain of this, check the phase diagram again.] As a result of this 0.8% C composition, we pay attention to Sketch 11 for a 1080 steel from here on.

After 3-b, the microstructure is partially bainite and partially untransformed γ.

After 3-c, part of the untransformed γ forms M; the already formed bainite remains.

After 3-d, martensite changes to tempered martensite. This is essentially indistinguishable from bainite. The residual austenite also produces the same product. Thus 97% of the steel is a fine dispersion of carbides in a ferrite matrix. A network of carbide continues to be present as 3% of the steel.

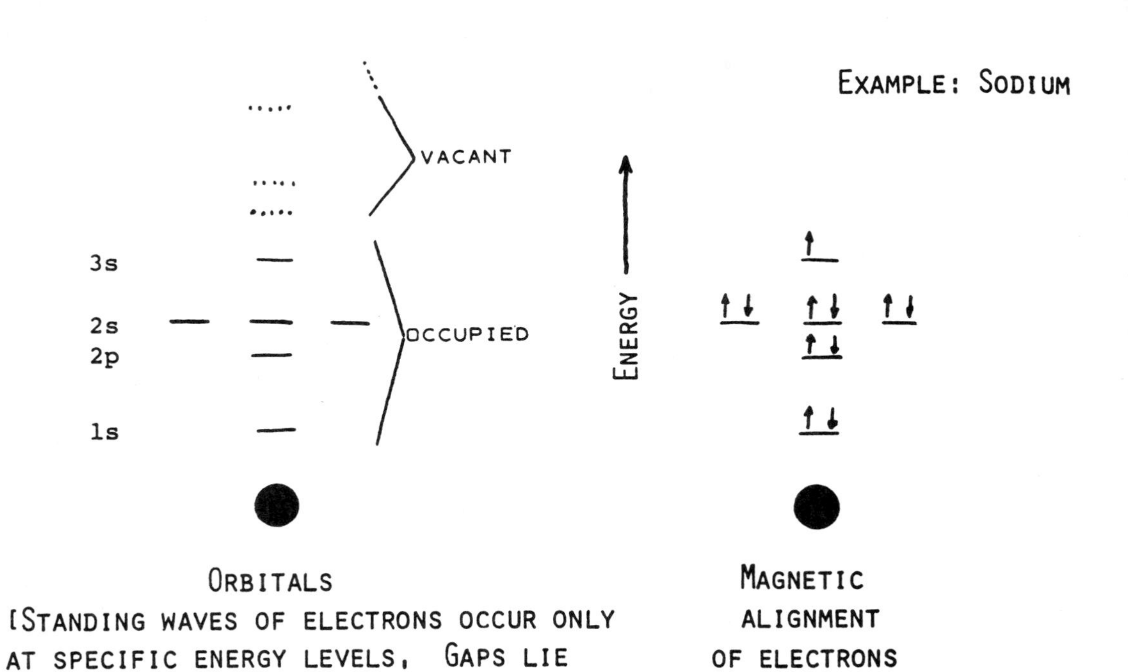

ORBITALS
[STANDING WAVES OF ELECTRONS OCCUR ONLY AT SPECIFIC ENERGY LEVELS. GAPS LIE BETWEEN THESE LEVELS.]
A

MAGNETIC ALIGNMENT OF ELECTRONS
B

VI

METALLIC

CONDUCTIVITY

1. **Standing waves, energy levels, and energy gaps** (opposite page). Physicists tell us the electrons that are associated with individual atoms move in a wavelike manner. These are *standing waves* and are called *orbitals.* As shown in (A) for sodium, each orbital has a characteristic energy. Only these energies support standing waves. Of equal importance to us are the *energy gaps* between these levels. Electrons cannot possess intermediate energies, because the corresponding standing wave would not be possible.

   An electron in a lower energy level, or "state", may be excited to an unfilled higher level; however, to do this, energy must be supplied for the electron to "jump the gap".

   No more than two electrons can be associated with each of these energy levels. These two must have opposite magnetic moments, or "spins". The spin alignments are shown in (B) for an atom of sodium, which has eleven electrons.

ORBITALS AND BANDS

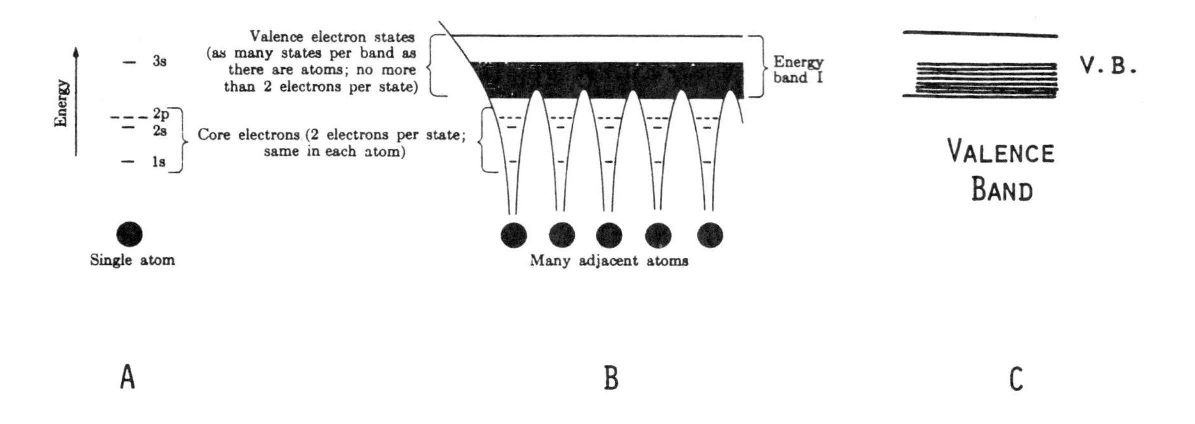

2. **Energy bands.** Rather than single atoms, we find a multitude of atoms in any engineering solid of useable size. The valence electrons of these atoms interact and become delocalized. Each time we add another atom, another possible wave form is developed for the valence electrons. With n atoms, there are n possible standing wave forms. The energies of these numerous standing waves, or states, are distinct, but closely associated. Consequently, we speak of an *energy band* . As shown in (B), the valence orbital of sodium has expanded into a valence band.

   Now, since each standing wave can contain two electrons (of opposite spin alignment), a solid of n atoms with n states can accomodate $2n$ electrons. Sodium has only one valence electron per atom, so its valence band is only half full. In fact, we define *metals* as being those materials with only *partially filled valence bands* (V.B.). The simplified sketch, which we will use, is shown in (C).

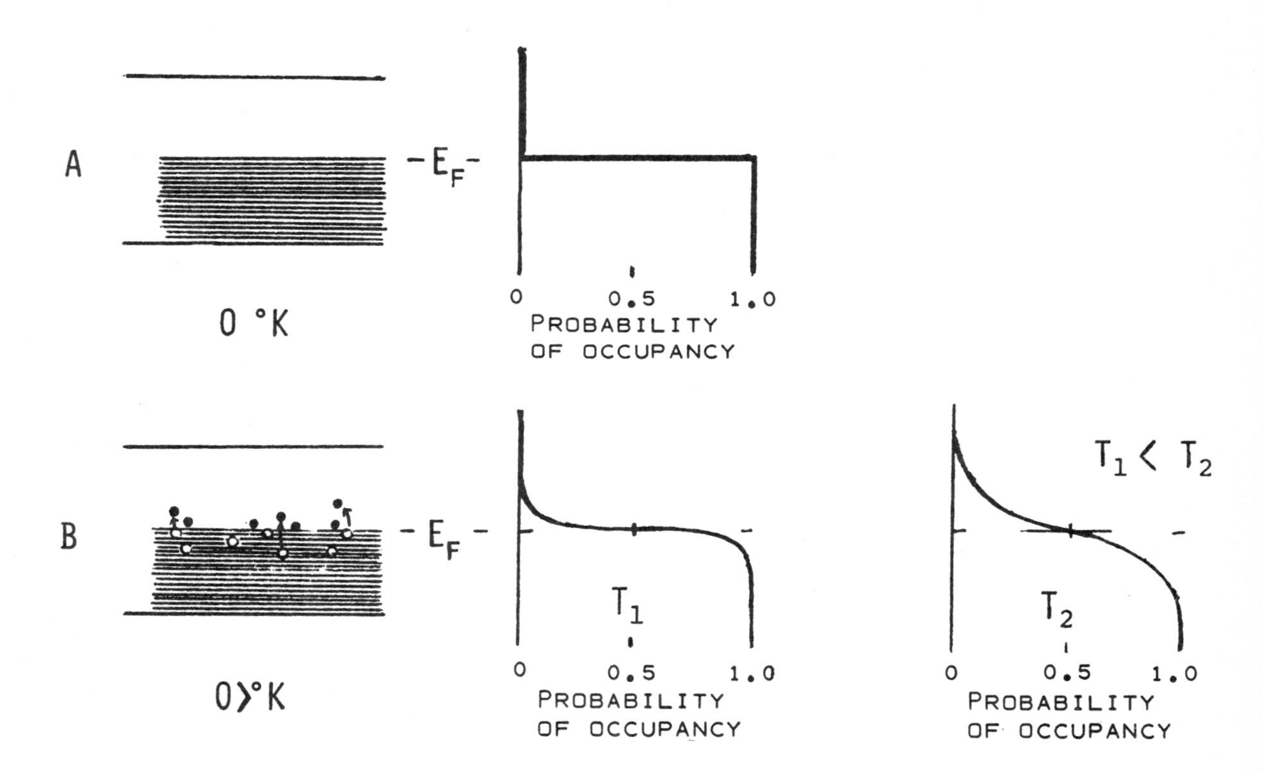

3. **Fermi energy.** If an energy band is not full, the electrons naturally occupy the lowest states (A). For sodium, and at 0°K, every state in the lower half of the band is occupied (with two electrons each). None of the states are occupied in the upper half of the band. The "sea level" of this "electron ocean" is called the *Fermi energy*, E_f.

   With increased temperature, or with any other energy source, some of the electrons can be activated into the upper part of the band above this Fermi energy (B). This is easy to do, since within the partially filled band of a metal, there are plenty of vacant states at only slightly higher energies. This leaves *electron holes* below the Fermi energy level.

   The states within the energy band have their probability of occupancy changed as the temperature is increased.

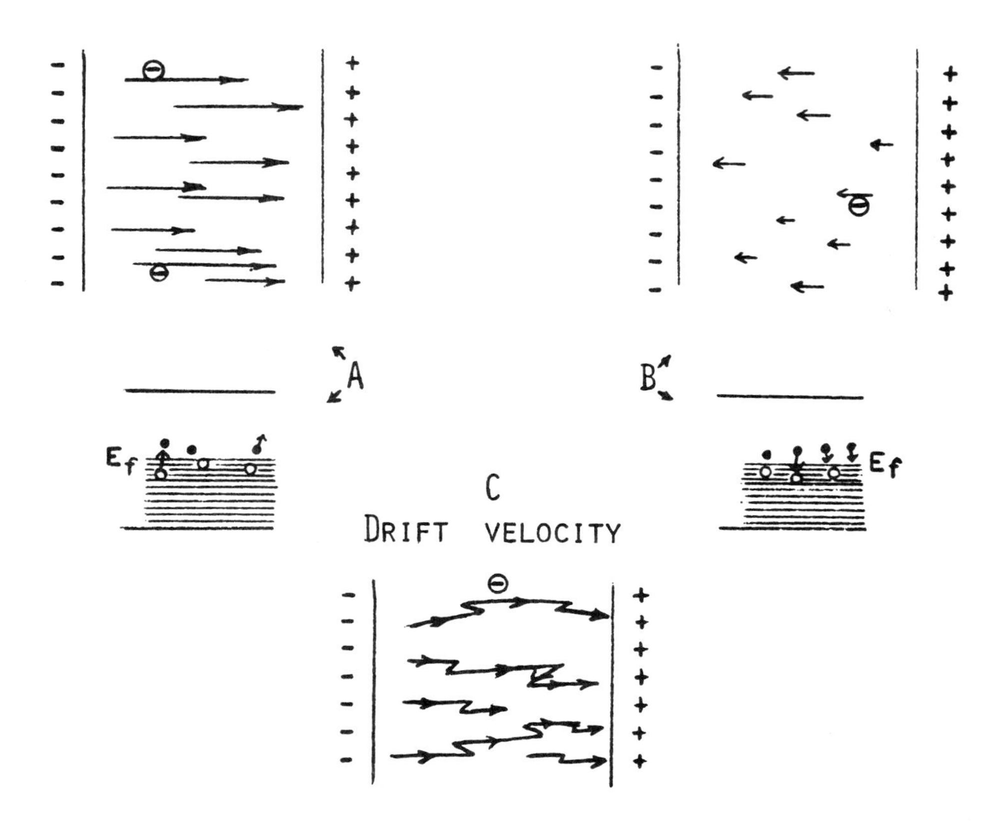

4. **Drift velocity.** The vacant part of the valence band permits conduction. As an electron moves in the direction toward the positive electrode in (A), it can receive additional energy (by jumping to incrementally high states). That increased energy appears as added momentum and a greater velocity in that direction.

If some of the electrons are above the Fermi energy, there must be *electron holes* below this "mean sea level." This permits an electron moving away from the positive electrode to lose velocity by dropping into a lower energy state (B). The overall effect is that there is a net movement of electrons toward the positive electrode. They have accelerated movements in that direction. Then if deflected or reflected, they have decelerated movements in the opposite direction (C). Specifically, charge is transported and conduction occurs by a net *drift* of electrons. This would not be possible if the valence band were full. We shall be interested in the *drift velocity* of the charges being transported.

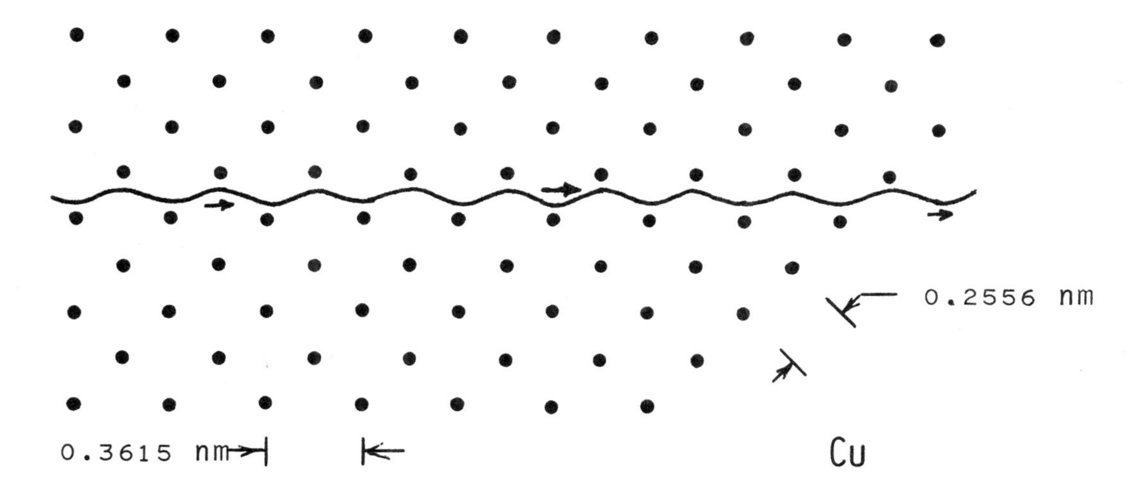

5. **Mean free path.** Since electrons move through solids in a wavelike manner, we find that they obey the laws of waves. One of these laws recognizes that waves travel furthest without reflection or deflection, if they travel through a very regularly arranged, highly ordered material. A crystal lattice of atoms has those features. In fcc copper, for example, atoms repeat every 0.2556 nm in the <110> directions, and 0.3615 nm in the <100> directions over thousands of atomic distances. We call this the *mean-free-path*. For the mean-free-path to be long, the crystal must be perfectly constructed. (The wave shown above is schematic only.)

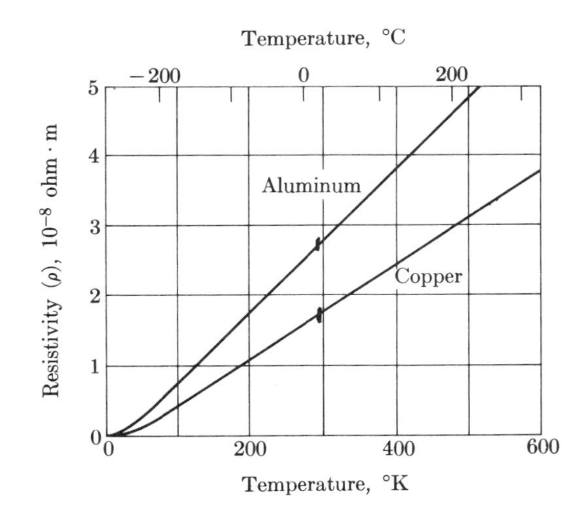

6. **Temperature-dependent resistivity.** Atoms are not static within crystals, but continually vibrate from thermal agitation. Thus, the wavelike electrons do not "see" a perfectly ordered, absolutely ordered structure. They are deflected and reflected. Their mean- free-path is shortened.

As the temperature increases, there is still more thermal disorder, and still more interference with electron movements. Therefore, the net flow of electrons and current is reduced. As shown here, the *temperature-dependent resistivity* of a metal increases at higher temperatures. The slope of the curve is the *thermal resistivity coefficient.* (Semiconductors and insulators are affected by other factors. Therefore, we can not generalize the above principles beyond metals.)

RESISITIVITY (AND CONDUCTIVITY) VS SOLID SOLUTION

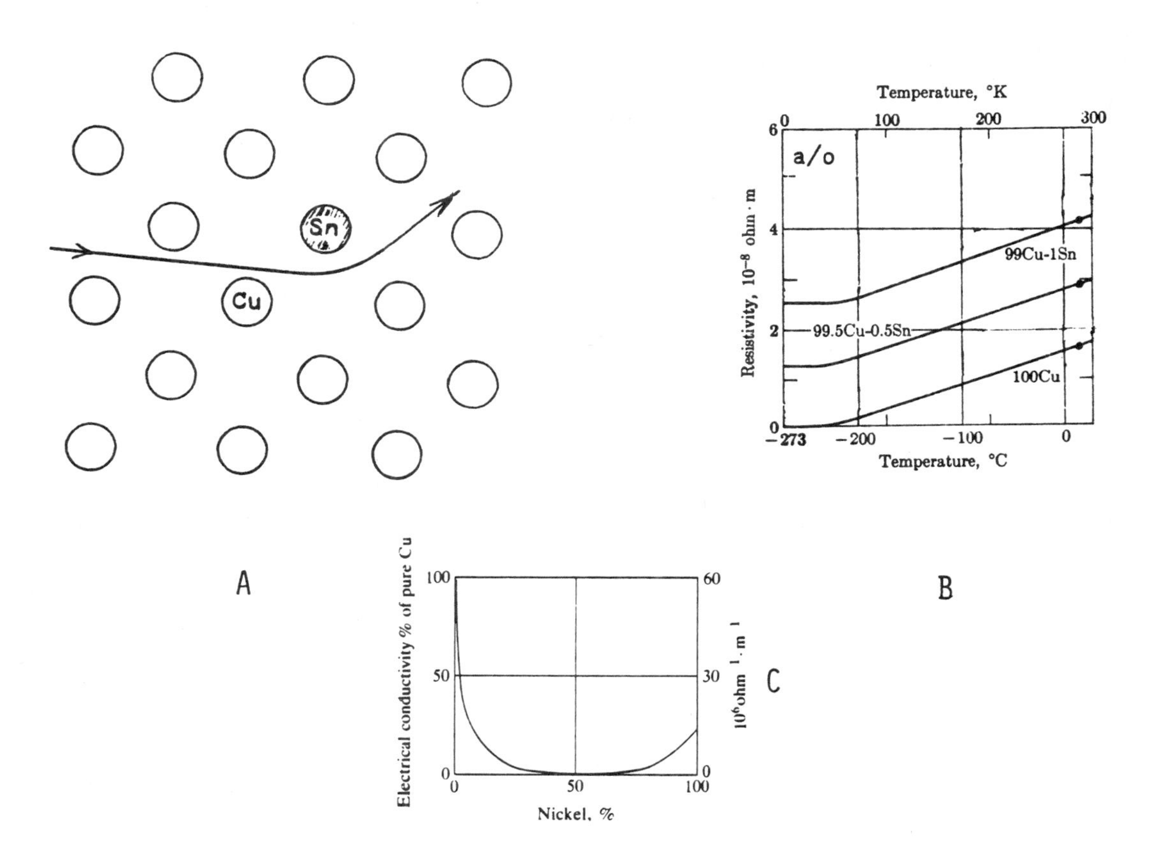

7. **Solution resistivity.** A solute atom introduces an irregularity within the crystal. For example, a tin atom has 60 protons, while a copper atom has 29 protons. If an electron that is moving through a copper crystal passes close to an impurity tin atom, its path can be sharply deflected, as sketched in (A). As the numbers of tin atoms in copper are increased to one atom percent (1.7 w/o), the resistivity at 20°C more than doubles, and the conductivity is cut in half. This is shown in (B). We call it, *solution resistivity.*

   In (C) we see the effect on conductivity when copper and nickel form solid solutions. Nickel lowers the conductivity of copper; and copper lowers the conductivity of nickel.

RESISTIVITY vs PLASTIC STRAIN

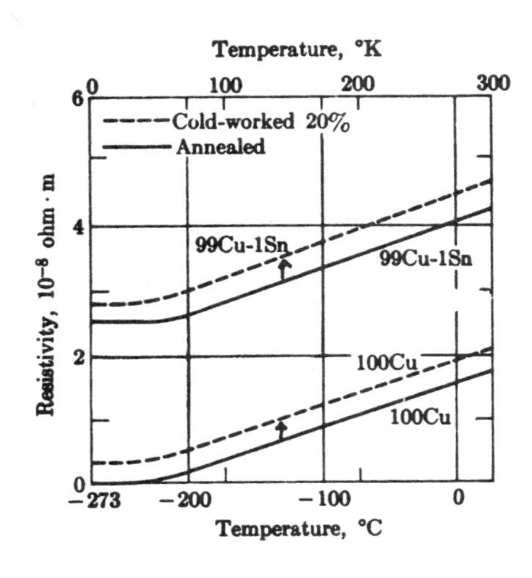

COPPER ALLOYS

A

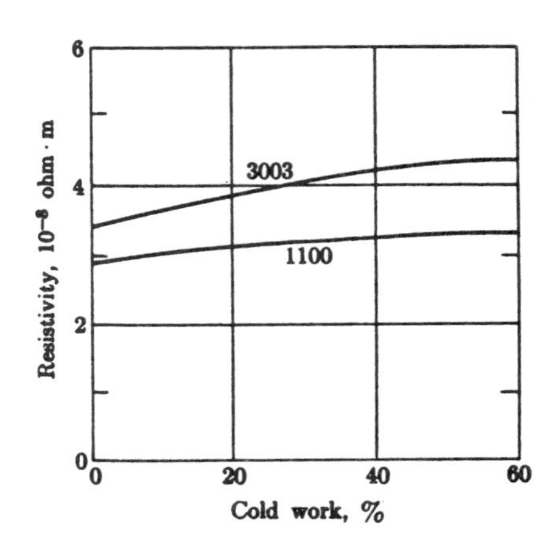

ALUMINUM ALLOYS

B

8. **Deformation resistivity.** A third type of irregularity is developed within a metal when it is plastically strained. Dislocations are introduced. They interfere with electron movements. Thus, in (A) we see added resistivity beyond that brought about by thermal agitation and by alloying elements. We call this *deformation resistivity*, or *strain damage.*

Part (B) shows the increase in resistivity with the strain that accompanies the cold working of aluminum and an aluminum alloy. The former (1100) is a commercially pure metal, such as used in pots and pans. When annealed so that the dislocations are removed, we have material used by power companies for electrical conductors. The 3003 alloy contains 1.2 w/o manganese to make it stronger. This also makes it more resistive, as shown here.

	LOW RESISTIVITY (HIGH CONDUCTIVITY)	HIGH RESISTIVITY (LOW CONDUCTIVITY)
TEMPERATURE	LOW	HIGH
SOLID SOLUTION	NONE (PURE)	ALLOYS
STRAIN	ANNEALED	DEFORMED

9. **Summary (metallic conductivity).** Let us review the information that we have on metallic conductivity. *Metals* are conductive because they have *unfilled valence bands.* This permits electrons to develop a net movement along a potential gradient and to conduct electricity.

   Electrons move in a *wavelike* manner. Therefore, electrons move more readily in more perfect structures, and have higher resistivities in less perfect structures. As a result, the resistivity of metals increases: 1) with increased *temperature*, 2) with increased *solid solution*, and 3) with plastic *deformation.*

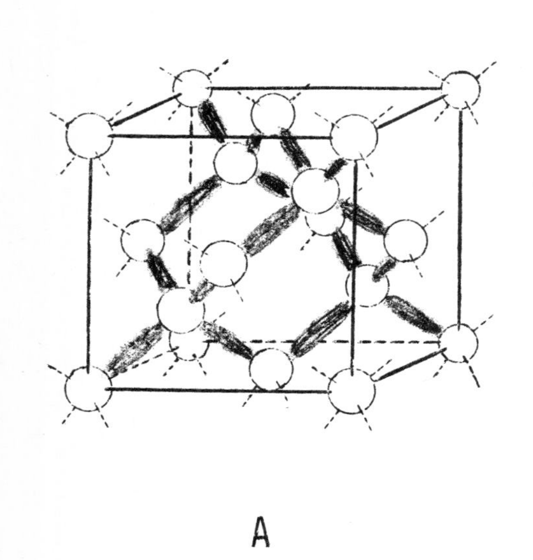

A

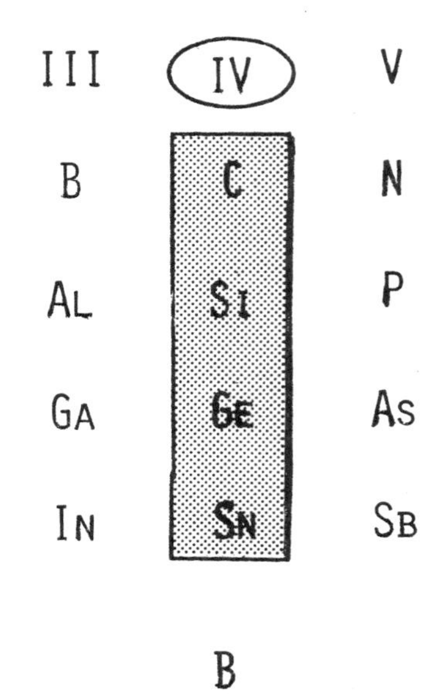

B

VII

SEMICONDUCTIVITY

1. **Basic semiconductor structures** (opposite). This sketch shows the most common crystal structure possessed by semiconductors. Other crystal structures exist; however, this one, plus the variant on the next page, will serve our needs.

   All atoms are the same in this structure, which is the structure of *diamond*, silicon, germanium, and the low-temperature form of tin (gray). Observe that in each case, the atoms of C, Si, Ge, and Sn reside in Group IVA of the periodic table. This means that they all have four valence electrons. Since each atom has four neighbors, each of the bonds contains a pair of electrons. These pairs produce covalent bonds.

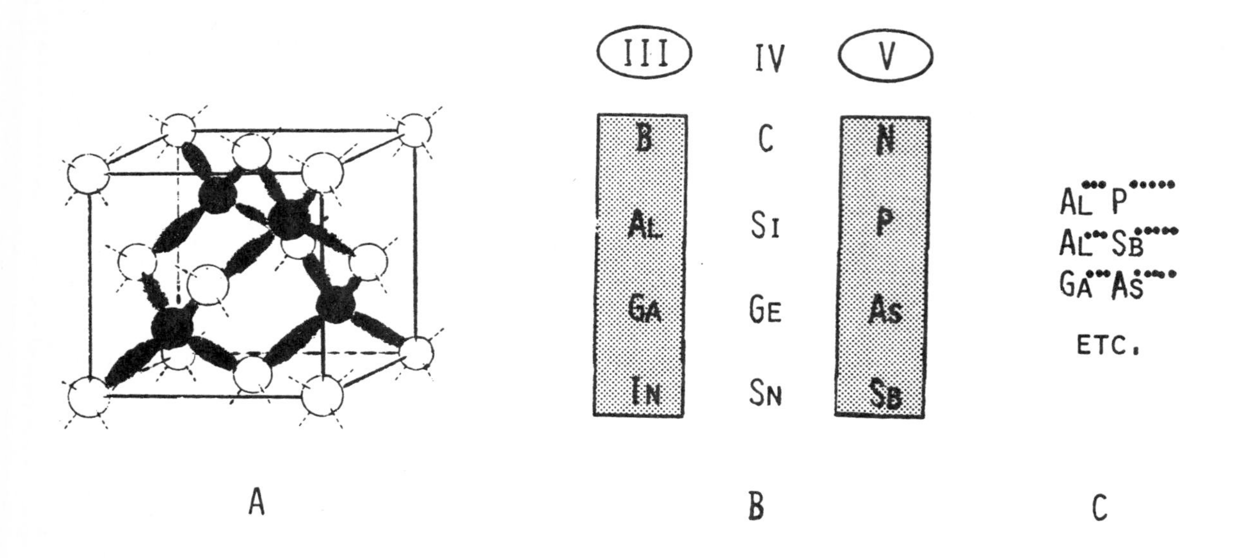

2. **III-V compound semiconductors.** This structure has the same pattern as that of diamond in the previous sketch, except that adjacent atoms are unlike. Alternate atoms are of one kind; intervening atoms are a second kind. Zinc sulfide is the prototype for this structure. We will note that it is also the structure of many *semiconducting compounds* such as AlP, GaAs, and InSb. Observe that these are III-V compounds, i.e., one element is from Group III of the periodic table, and the other is from Group V. Thus , these compounds average four valence electrons per atom. Since each atom has four neighbors, each of the bonds contains a pair of electrons. As in diamond, these are covalent bonds.

Our consideration of semiconductors will apply to the diamond structure on the previous page, and to the above ZnS structure .

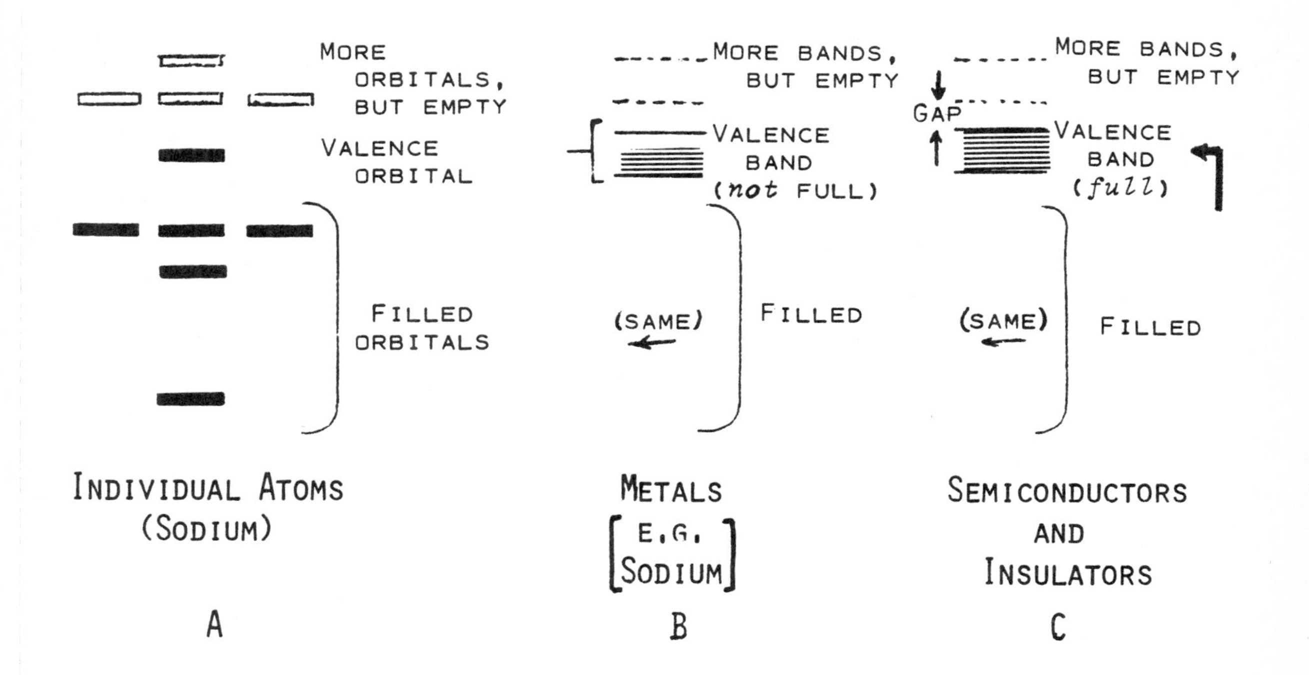

3. **Review of conductivity principles.** Before examining how semiconductors transport charge, let us briefly review orbitals, bands, and their related energy gaps.

Sketch (A) shows the energies of orbitals for electrons associated with *individual* sodium atoms. Each is the energy of the standing waves for a pair of electrons (with opposite magnetic orientation).

Of course, any useable engineering solid has a multitude of atoms. The waves for the valence electrons of these numerous atoms interact to form a *valence band* as shown in (B). In *metals,* such as sodium shown here, the valence band is only partially filled, so that only half of the energy "states" are occupied.

In *semiconductors* and *insulators,* every state of the valence band is completely filled, as shown in (C) for silicon. Every standing wave is utilized, each containing a pair of electrons (of opposite magnetic alignment).

There is a gap above the last filled valence band. Beyond this *forbidden gap,* there are empty bands. This gap, along with the underlying valence band (VB) and the overlying band, called the conduction band (CB), will receive our attention in the balance of this study set.

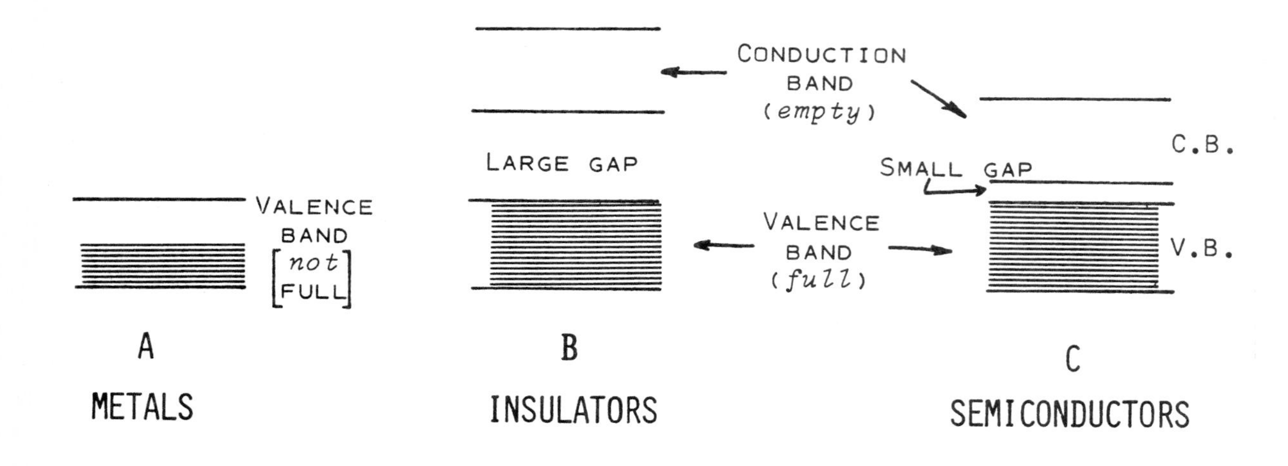

4. **Metals, insulators, and semiconductors.** Electron energy bands, and the width of the *forbidden energy gap* between them, play an important role in the electrical properties of materials.

(A) As indicated on the previous page, *metals* have unfilled energy bands. They are discussed under "Metallic Conductivity" in Study Set VI.

(B) *Insulators*, such as diamond and AL_2O_3, have very wide forbidden energy gaps. We shall observe that this is a barrier to electrical conductivity.

(C) *Semiconductors*, such as silicon (Si), germanium (Ge), and gallium arsenide (GaAs) have relatively narrow energy gaps. We shall consider these on subsequent pages.

The band below the energy gap of insulators and semiconductors is called the *valence band*, V.B. The band above the energy gap is the *conduction band*, C..B.

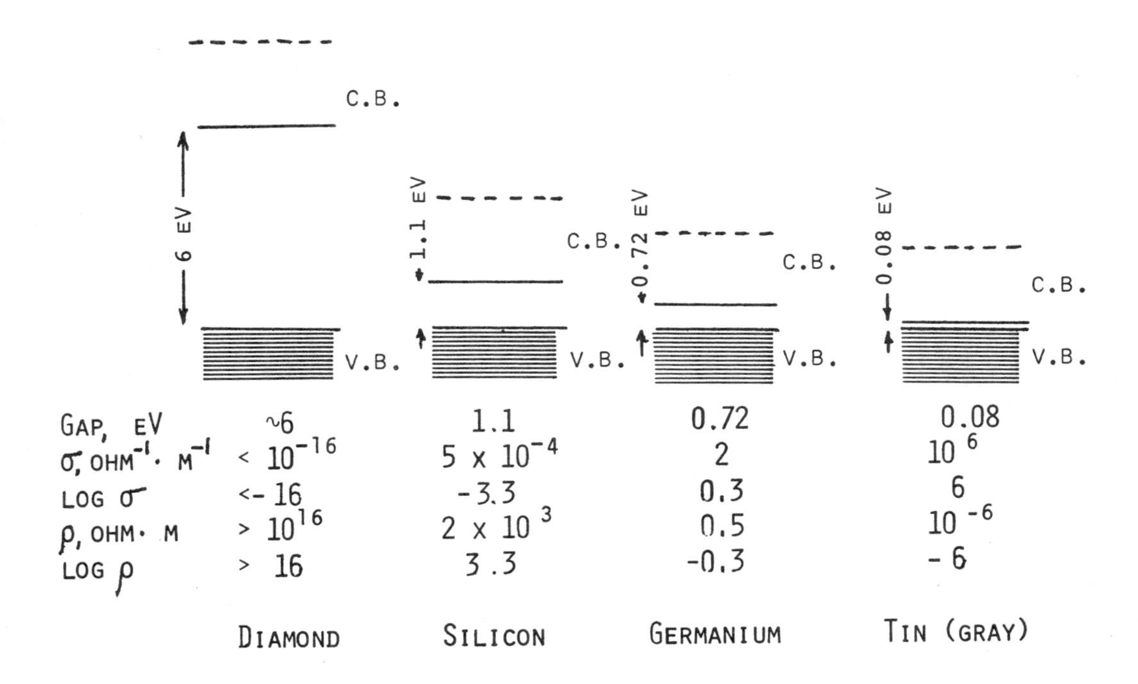

5. **Gap size *vs.* conductivity.** Since diamond, silicon, germanium, and gray tin are Group IV elements, and all have the same structure, we can compare them directly. Their *energy gaps*, E_g, are shown schematically in the above sketches.

Diamond, with $E_g = 6$ eV, has a high resisivity of 10^{16} ohm·m, which makes it an excellent insulator.

Silicon, with $E_g = 1.1$ eV, has a low resistivity of 2×10^3 ohm·m at 20°C. Silicon does not pass as a satisfactory insulator.

Germanium has a relatively high conductivity ($2\ \text{ohm}^{-1}\cdot\text{m}^{-1}$), and still lower resistivity (0.5 ohm·m), because its energy gap is only 0.7 eV. Incidently, 1 eV is 0.16×10^{-18} J, because each electron carries a charge of 0.16×10^{-18} coulombs.

Gray tin is almost metallic. (White tin is.) The energy gap of gray tin is < 0.1 eV, which gives it a conductivity of $10^6\ \text{ohm}^{-1}\cdot\text{m}^{-1}$, and a resistivity of only 10^{-6} ohm·m.

Observe that there is a relationship between the logarithm of the conductivity (or resistivity) and the size of the energy gap for these materials (A).

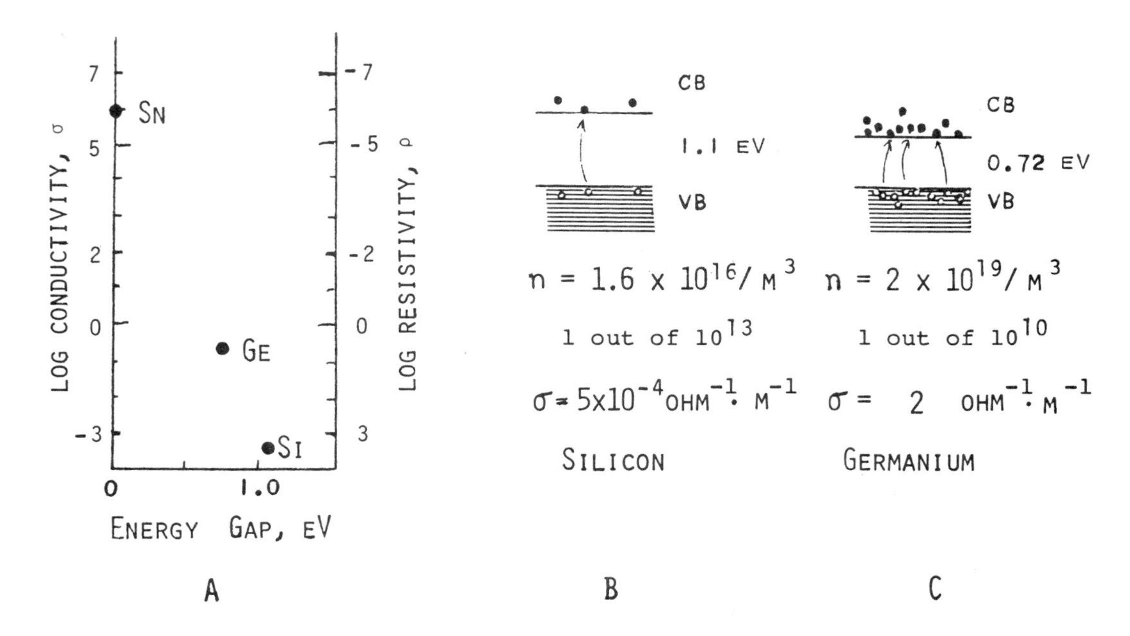

6. **Gap size *vs.* conductivity** (con't). The data on the previous age should be examined more, since we may legitimately ask why the size of the energy gap affects these properties. Let's start with an insulator such as diamond.

The *valence band* of diamond is full. Every state is occupied; every standing wave is utilized. Within the band, electrons cannot receive added energy for greater momentum and velocity to respond to the electric field. Nor can they decrease their energy as they move toward the negative electrode. Within the band there is no opportunity for a net movement of charge toward the positive electrode.

In order to gain more momentum or energy for conductivity, the electron would have to "jump the gap" to the *conduction band.* In diamond, that *energy gap* is 6 eV. This is much more than the average energy of a valence electron, which is only 0.025 eV at 20°C. A few (only 1 out of 30×10^{21}) will have enough energy to jump the gap; but, this number is insufficient to give significant conductivity; in fact, it possesses a resistivity of more than 10^{16} $ohm^{-1} \cdot m^{-1}$, which we encountered on the previous page.

In silicon, with an energy gap of 1.1 eV, (B), one out of every 10^{13} electrons has sufficient energy to "jump the gap" at 20°C. This number provides for a conductivity of 5×10^{-4} $ohm^{-1} \cdot m^{-1}$, a low but readily measured and useable quantity. In germanium, (C), the fraction, and the resulting conductivity are still higher.

The electrons that jump the gap can transport charge, because there is room in the conduction band for them to be accelerated and gain energy and velocity as they move toward the positive electrode. The whole of the conduction band becomes an "open highway" for charge transfer.

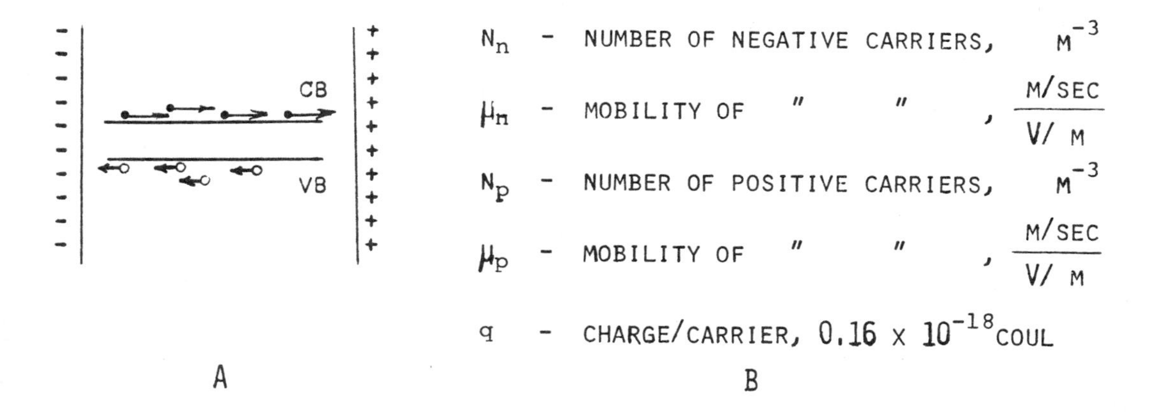

7. **Charge carriers (electrons and holes), and their mobilities.** To date, we have discussed only the electrons that have jumped across the gap from the valence band to the conduction band. Those electrons contribute to the conductivity because they carry a charge toward the positive electrode. The *electrons* are negative *carriers* and have high mobility in the unobstructed *conduction band.*

For every electron that crosses the energy gap, an *electron hole* exists in the *valence band.* This also permits conductivity . A crude analogy is to say that "the traffic jam in the filled valence band has been relieved." The electron holes provide openings into which other electrons can move. In an electric field, these holes move toward the negative electrode as the electrons move toward the positive electrode. We therefore say the the electron hole carries a *positive* charge. Mathematically,

$$\sigma = n_n q \mu_n + n_p q \mu_p .$$

The *mobilities,* μ_n, of these positive carriers (holes) in the valence band are never as great as the *mobility,* μ_n, of an electron in the conduction band. This is plausible because the conduction band has many vacant states into which the accelerating electrons can advance.

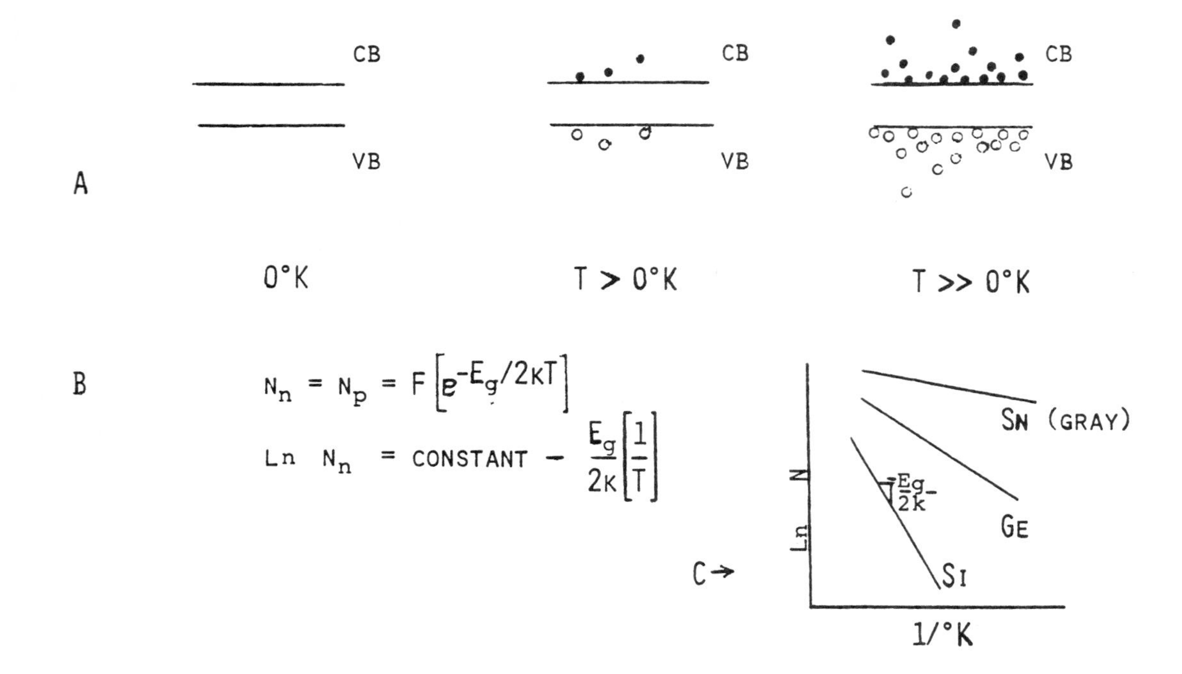

8. **Effect of temperature on the number of charge carriers.** Experiments show that more and more electrons "jump the gap" in silicon as the temperature is raised (A). This is because electrons, like atoms, pick up added thermal energy when we supply heat.

The increase in the *number* of electrons in the conduction band does not vary linearly with temperature ; rather, *exponentially*! As shown in (B), the number of negative carriers, n_n, is an exponential function of the size of the energy gap, E_g, and a function of the reciprocal of absolute temperature, T. We will not derive this, but will note that log n_n is linearly related to $1/T$. This gives us the curve shown in (C), where the slope is $E_g/2k$, and k is the widely encountered Boltzmann's constant equaling 86.1 $\times 10^{-6}$ eV/°K. Observe that not only the numbers of negative carriers increase from Si, to Ge, to gray Sn, but also the slope of the curves varies significantly. We expect this, since E_g decreases through that sequence.

A

		Silicon	Germanium	Units
Energy gap,	E_g	1.1	0.72	eV
$n_n = n_p$ *		2×10^{16}	2×10^{19}	m^{-3}
Charge, q		0.16×10^{-18}	0.16×10^{-18}	coul
Mobility				
electrons,	μ_n	0.19	0.36	$\frac{m/s}{volt/m}$
holes,	μ_p	0.0425	0.23	"

* At 20°C

B

$$\sigma = n_n q \mu_n + n_p q \mu_p$$

$$\sigma_{Ge@20°C} = (2 \times 10^{19}/m^3)(0.16 \times 10^{-18}\,coul)\left(0.36 + 0.23 \frac{m/s}{V/m}\right)$$

$$= 2\ ohm^{-1} \cdot m^{-1}$$

C

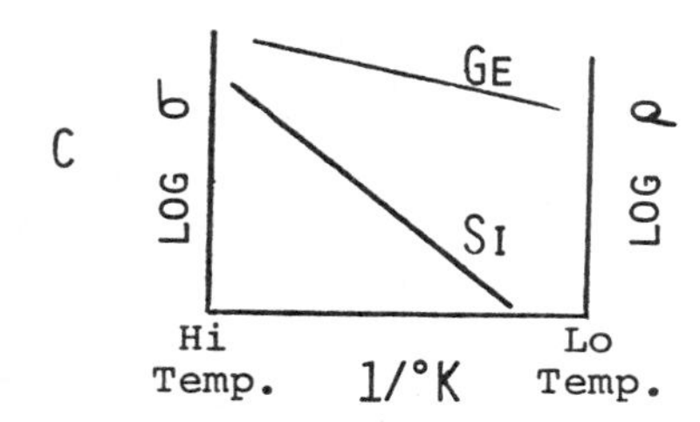

9. Intrinsic semiconduction. Charge is carried by electrons (*n*-type carriers) in the conduction band, and by electron holes (*p*-type carriers) in the valence band. The conductivity, which is intrinsic to the pure semiconductors that we have been examining, arises from equal numbers of *n*-type and *p*-type carriers., i.e. $n_n = n_p$. Both types possess the same charge, $q = 0.16 \times 10^{-18}$. As noted previously, the mobilities of the two types are not equal, $\mu_n > \mu_p$.

With the information of (A), we can calculate the *conductivity*, σ, of intrinsic semiconductors, because

$$\sigma = n_n q \mu_n + n_p q \mu_p \, .$$

For germanium at 20°C, where $n_n = n_p = 2 \times 10^{19}/m^3$, we calculate in (B) that $\sigma = 2\ ohm^{-1} \cdot m^{-1}$.

We saw in the previous sketch that n increases exponentially. Therefore, the conductivity also increases exponentially. (The charge, q, is constant , and the mobility values, μ, change only linearly with temperature.) Thus, we have nearly a straight line when we graph log σ *vs.* reciprocal absolute temperature, $1/T$. This is an Arrhenius plot.

IV	V
C	N
Si	P
Ge	As
Sn	Sb

0.1 PPM = 1 P PER 10^7 Si

5×10^{28} Si/ $m^3 \Rightarrow 5 \times 10^{21}$ P/ m^3

WITH 5×10^{21} P/ m^3

$n = 5 \times 10^{21}$ CARRIERS/ m^3 ADDED

VS $n_n = n_p \cong 10^{16}$ IN PURE Si

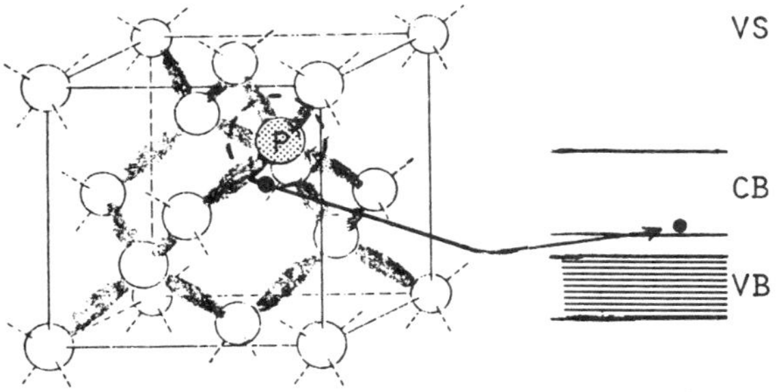

CB

VB

$q = 0.16 \times 10^{-18}$ COUL

$\mu_n = 0.19$ m^2/VOLT·S

$\mu_p = 0.0425$ "

10. **Extrinsic semiconduction (*n*-type).** Semiconductors with compositional modifications are called *extrinsic semiconductors.* Most common among these is silicon that has been doped with aluminum, boron, phosphorus, or arsenic. We will observe marked and useful differences in the characteristics of these extrinsic semiconductors as compared to *intrinsic* (pure) semiconductors.

We will consider the phosphorus doping of silicon first. Phosphorus is a Group V element. Assume we add only 0.1 ppm of P to Si, an extremely small fraction. There are, however, 5×10^{28} Si atoms/m^3. This means adding 5×10^{21} P atoms/m^3; not an insignificant number.

Since phosphorus is a Group V element, it has five valence electrons ,-- one more than the basic four for silicon. This fifth electron does not reside in the already filled valence band, and can easily be pulled away from the P atom and *donated* to conduction band, CB. In fact at 20°C, essentially all of the phosphorus atoms are ionized, losing this fifth electron to the conduction band.

With 5×10^{21} P/m^3, there can be 5×10^{21} negative carriers per m^3, i.e., $n_n = 5 \times 10^{21}$ m^3. Recall that at 20°C, intrinsic silicon has only 10^{16} electrons per m^3 that have jumped the gap. Thus, the number of carriers added by this small addition of phosphorus far exceeds the number of intrinsic carriers in silicon.

Calculate the conductivity for the above extrinsic semiconductor and check yourself on the next page. [The *mobilities* of electrons and electron holes in silicon are 0.19 m^2/V·s and 0.0425 m^2/V·s, respectively.]

SOLUTION: (n-type Si, previous page)

$$\sigma = (nq\mu)_{extr} + (n_n q\mu_n)_{intr} + (n_p q\mu_p)_{intr}$$

$$= (5 \times 10^{21}\ \mathrm{m^3})(0.16 \times 10^{-18}\ \mathrm{coul})(0.19\ \mathrm{m^2/volt \cdot sec}) + \text{negligible}$$

$$= 150\ \mathrm{ohm^{-1} \cdot m^{-1}}$$

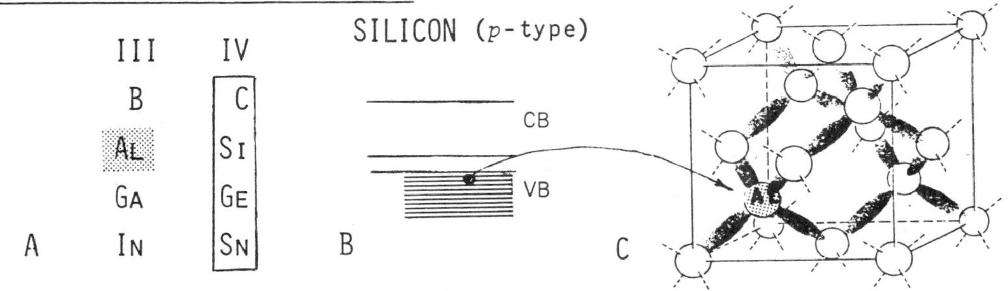

0.1 ppm Al = 1 Al per 10^7 Si

5×10^{28} Si/ $\mathrm{m^3}$ give 5×10^{21} Al/ $\mathrm{m^3}$

$$n_p = 5 \times 10^{21}\ /\ \mathrm{m^3}$$

$$\sigma_{extr} = (5 \times 10^{21}/\mathrm{m^3})(0.16 \times 10^{-18}\ \mathrm{coul})(.0425\ \mathrm{m^2/V \cdot sec}) = 35\ \mathrm{ohm^{-1} \cdot m^{-1}}$$

11. **Extrinsic semiconduction (*p*-type).** The calculation of the previous problem shows the conductivity of phosphorus-doped silicon is 150 $\mathrm{ohm^{-1} \cdot m^{-1}}$. Since the extrinsic conductivity in this silicon is predominantly by electrons in the conduction band, we call this an *n-type semiconductor.*

    Now let us consider the addition of a like amount of aluminum to silicon, *viz* ., 0.1 ppm, or 5×10^{21} Al atoms/$\mathrm{m^3}$. Aluminum is a Group III element; therefore, it has only three valence electrons per atom, as compared to four for silicon. Consequently, the bonds around the atom are short an electron. As a result the aluminum atoms will *accept* electrons from the valence band, leaving holes as indicated in (B).

    With 5×10^{21} Al atoms/$\mathrm{m^3}$, there are 5×10^{21} electron holes/$\mathrm{m^3}$ in the valence band. These are *p*-type carriers, and the Al-doped silicon is a *p-type semiconductor.* The extrinsic conductivity is the product of $5 \times 10^{21}/\mathrm{m^3}$, 0.16×10^{-18} coulombs, and 0.0425 $\mathrm{m^2/V \cdot s}$ (which is the mobility of the holes in the valence band of silicon).

    Other *p*-type dopants include boron, gallium, and indium,-- all Group III elements. You cite 3 or 4 *n*-type dopants for silicon.

QUIZ CHECKS *

Chapter 1 INTRODUCTION to MATERIALS SCIENCE and ENGINEERING

1 A stone, bronze, iron
1 B iron
1 C materials, energy
1 D lose
share
compounds
1 E 100 MPa
1 F force, area
deformation, length
newton
1 G lower
stress

Chapter 2 ATOMIC BONDING and COORDINATION

2A (b), (c)
2B 0.16×10^{-18}
coulomb, C
2C intra-
inter-
2D (b), (d), (e)
2E mers
polymer
2F chlorine
benzene ring
vinyl alcohol
$-CH_3$
2G two
one and two
2H mers
2I (a), (c)
2J (b), (c)

Chapter 3 CRYSTALS (ATOMIC ORDER)

3A (b), (d)
3B orthorhombic, triclinic
3C (a)
3D four
3E fcc
hcp or fcc
3F 0.1, 0.7, 0.95
3G (c), (e)
3H (e): $^1/_2 + n$, $^1/_2$, $^1/_2 + 2n$
3I (c)
3J $(\bar{1}01)$ or $(10\bar{1})$
3K (a), (b), (c), (d), (e)
3L 1.8 /nm
1.12 nm
3M 0.209 nm
0.128 nm
0.181 nm
3N (a), (b)

---

* Answers to the Quiz Samples of pp, 1-23.

Chapter 4 DISORDER in SOLID PHASES

4A (b), (c)
4B parallel
energy
4C surfaces
grains or phases
4D (a), (c)
4E melting
supercooled
4F degree of polymerization
4G mass or number
mass
4H 40%
4I copolymer
4J isotactic
atactic
4K (a), (b)
4L (e): cross-linking
4M H
CH_3
chloroprene
isoprene
4N amorphous
glass
supercooled
4O free space
4P increases
4Q (a), (c)
4R lower
lower
4S 540
4T nonstoichiometric

Chapter 5 PHASE EQUILIBRIUM

5A solution
phases
5B solubility limit
low
5C (e)
(b)
(e)
5D face
zinc
bronze
5E (b), (c)
5F (a), (c)
5G two-phase, (c)
(b)
5H peritectic
5I α, γ, and $\overline{C}$

Chapter 6 REACTION RATES

6A (c), (e), (f)
6B (a), (b)
6C (d) 400 to 450°C
6D increases
6E impurities, imperfections, surfaces, "seeds"
6F increases
smaller
6G (e) log n vs. $°K^{-1}$
6H concentration gradient
6I solute
lower
lower
6J m^2/s
6K (e) log D vs. $°K^{-1}$

Chapter 7 MICROSTRUCTURES

7A size, shape, orientation
7B amounts, distributions
7C (b), (c)
7D 4, 2
7E curvature, large
7F increases, increases
7G (a), (c)
(b)
7H (b), (e)
(a), (d), (f)
7I 727°C, 0.77 (or ~ 0.8)
$\alpha + \overline{C} \xrightarrow{\text{heating}} \gamma$

7J pearlite, $\alpha + \overline{C}$
7K carbide
ferrite
7L (b), (d), (e)
(a), (f), (g)
7M (a)
(c)
7N (d): grain growth,
or recrystallization,
or martensite formation
7O (b), (d), (e), (f)

Chapter 8 DEFORMATION and FRACTURE

8A elongation, reduction of area
fracture
8B yield
elastic
8C ultimate
original
8D the area decreases before breaking
(necking occurs)
8E Poisson's, lateral
volume
8F tangent
8G hydrostatic pressure-to-($\Delta V/V$)
8H (b), (c), (f)

8I ductile, brittle
8J stress
energy
8K loss, ductility-transition
body
face
8L fine
8M size, shape
8N (b)
8O notch (or flaw)
tip (or end)
square root of crack depth

Chapter 9 SHAPING, STRENGTHENING, and TOUGHENING PROCESSES

9A (e)
9B (b)
9C drawing, rolling, forging,
extrusion,etc.
9D (d): 17%
9E (e)
9F (a)
9G (e): log t vs. K^{-1}
9H (d), (e), (f)
9I heat above, quench below
9J (c)
(a)
9K (b), (c), (d)
9L (b), (d), (e)
(a), (f), (g)
9M (a), (c)

9N (e): grain growth,
or recrystallization,
or martensite formation
9O (a), (c)
(b)
9P (b) (not a)
9Q (a), (b), (c)
9R (e): full anneal of steels
9S soaking, normalizing
9T thinner
9U (a), (c), (d)
9V indentation (or scratching)
hardenability
9W alloying elements
9X water
heat of vaporization

Chapter 10 POLYMERS and COMPOSITES

10A glass-transition temperature
more
10B relaxation
10C higher
10D elastic modulus
10E volume fraction
10F oriented
anisotropic
10G shear
10H tension
compression
10I (c)

Chapter 11 CONDUCTING MATERIALS

11A (a)
11B 0.16 ohm, 6×10^{16} el/ms
11C (c)
11D (e): 33×10^{6} $ohm^{-1} \cdot m^{-1}$
11E (d)
11F (c)
11G (a), (c)
11H metal
semiconductors and insulators
11I gap, insulator
11J (d): atomic weight,
or atomic number
11K conduction
valence
11L III and V, 4
heavier
11M absolute temperature
11N heat, light (radiation),
electrons
photo-, photons
11O -hole
recombination
11P extrinsic
intrinsic
11Q pressure gage
reduced
11R thermistor
temperature
11S rectifier, or LED
forward

Chapter 12 MAGNETIC PROPERTIES of METALS and CERAMICS

12A hard
coercive field
12B 4,5
3
12C boundaries
(a), (b), (c), (d)
12D (a), (b), (c)
12E 4-f, 6-f
opposite
12F 16
0.83 nm
12G induction/magnetic field
12H magnetization
unit volume
9.27×10^{-24}
zero
12I ferric, ferrous
five, four
12J antiferro-
ferri-
12K domains
boundaries
soft
12L *BH*
energy
12M dislocations
domain boundary
12N dislocations, second phases
pores, grain boundaries
12O Curie
thermal agitation destroys
the cooperative orientation

Chapter 13 DIELECTRIC and OPTICAL PROPERTIES

13A conductors
energy gap

13B volts/m
thickness
pores, impurities,
and structural imperfections

13C cracks, impurities, pores

13D (d): total dipole moment per volume,
or excess charge density

13E electronic, ionic, molecular
also interfacial (space charge)

13F electronic, ionic, molecular,
also interfacial (space charge)
electronic
molecular

13G dipole moment/unit volume
charge density

13H one
lower

13I glass transition
thermal agitation destroys the
orientation
electronic

13J electronic
molecular and atomic movements
are too sluggish

13K piezoelectric, strain

13L transducer

13M ferroelectric, piezoelectric

13N ferro- = piezo-
piezo- ≠ ferro-
some piezoelectric dipoles are
not reversible

13O remanent
coercive field

13P (d) greater velocity, or less electronic
polarization, or longer wavelength

13Q index of refraction

Chapter 14 PERFORMANCE of MATERIALS in SERVICE

14A (a), (b), (c), (d), (e)

14B (a), (b), (c), (d), (e)

14C anode, electrolyte

14D anode
electroplating

14E one, 25°C
(a), (c), (d)

14F $2\,H_2O + {}^1/_2\,O_2 + 4\,e^- \longrightarrow 4\,(OH)^-$

14G (a), (d), (e), (h), (j)

14H H_2

14I 1-molar
dilute

14J dilute

14K lower

14L anode
cathode
deformed
residual strain energy is present

14M anode ⇒ cathode
anode ⇒ cathode

14N passivated
activated

14O stress
brittle
static

14P oxide
glasses (ceramics)
paints (polymers)
metals
inhibitors

14Q anodic
into

14R zinc is anodic to iron
Al_2O_3 forms a protective film
the surface becomes passivated

14S fatigue
S-N curve
$^1/_4$

14T (c), (d), (e)

14U chemical reaction
and stress

14V (a)

14W 2
temperature, stress
grain size

14X diffusion
four

14Y decarburization

14Z carbonization (charring)

14AA scission

14AB (c)

14AC SiO_2, Al_2O_3, MgO
ZrO_2, Cr_2O_3

14AD strength and thermal
conductivity
thermal expansion
coefficient and
elastic modulus

14AE point
plastic deformation

14AF recovery
annealing (softening)

14AG scission
average

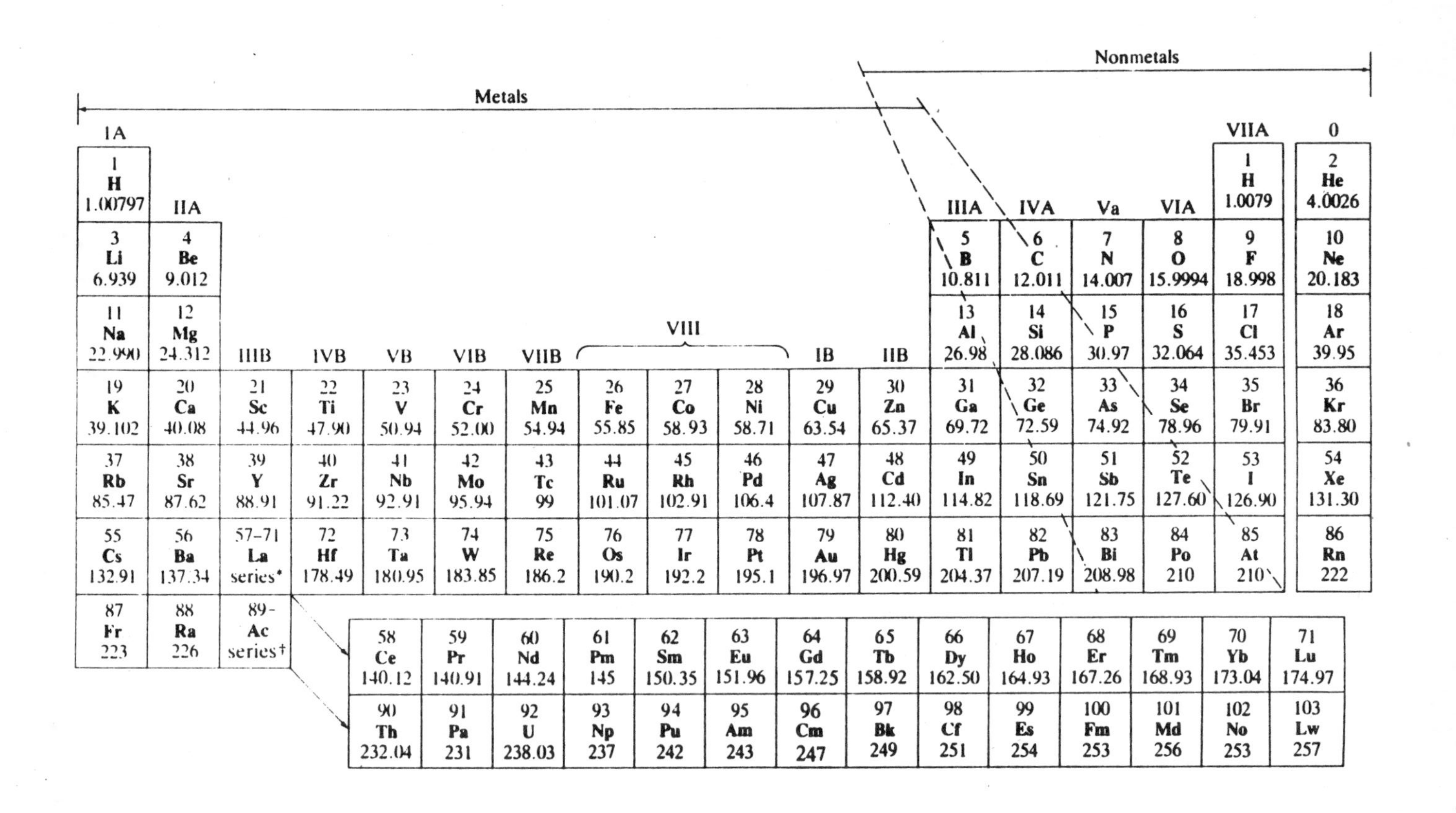

Metals | Nonmetals

IA	IIA	IIIB	IVB	VB	VIB	VIIB	VIII	VIII	VIII	IB	IIB	IIIA	IVA	Va	VIA	VIIA	0
1 **H** 1.00797																1 **H** 1.0079	2 **He** 4.0026
3 **Li** 6.939	4 **Be** 9.012											5 **B** 10.811	6 **C** 12.011	7 **N** 14.007	8 **O** 15.9994	9 **F** 18.998	10 **Ne** 20.183
11 **Na** 22.990	12 **Mg** 24.312											13 **Al** 26.98	14 **Si** 28.086	15 **P** 30.97	16 **S** 32.064	17 **Cl** 35.453	18 **Ar** 39.95
19 **K** 39.102	20 **Ca** 40.08	21 **Sc** 44.96	22 **Ti** 47.90	23 **V** 50.94	24 **Cr** 52.00	25 **Mn** 54.94	26 **Fe** 55.85	27 **Co** 58.93	28 **Ni** 58.71	29 **Cu** 63.54	30 **Zn** 65.37	31 **Ga** 69.72	32 **Ge** 72.59	33 **As** 74.92	34 **Se** 78.96	35 **Br** 79.91	36 **Kr** 83.80
37 **Rb** 85.47	38 **Sr** 87.62	39 **Y** 88.91	40 **Zr** 91.22	41 **Nb** 92.91	42 **Mo** 95.94	43 **Tc** 99	44 **Ru** 101.07	45 **Rh** 102.91	46 **Pd** 106.4	47 **Ag** 107.87	48 **Cd** 112.40	49 **In** 114.82	50 **Sn** 118.69	51 **Sb** 121.75	52 **Te** 127.60	53 **I** 126.90	54 **Xe** 131.30
55 **Cs** 132.91	56 **Ba** 137.34	57–71 **La** series*	72 **Hf** 178.49	73 **Ta** 180.95	74 **W** 183.85	75 **Re** 186.2	76 **Os** 190.2	77 **Ir** 192.2	78 **Pt** 195.1	79 **Au** 196.97	80 **Hg** 200.59	81 **Tl** 204.37	82 **Pb** 207.19	83 **Bi** 208.98	84 **Po** 210	85 **At** 210	86 **Rn** 222
87 **Fr** 223	88 **Ra** 226	89- **Ac** series†															

58 **Ce** 140.12	59 **Pr** 140.91	60 **Nd** 144.24	61 **Pm** 145	62 **Sm** 150.35	63 **Eu** 151.96	64 **Gd** 157.25	65 **Tb** 158.92	66 **Dy** 162.50	67 **Ho** 164.93	68 **Er** 167.26	69 **Tm** 168.93	70 **Yb** 173.04	71 **Lu** 174.97
90 **Th** 232.04	91 **Pa** 231	92 **U** 238.03	93 **Np** 237	94 **Pu** 242	95 **Am** 243	96 **Cm** 247	97 **Bk** 249	98 **Cf** 251	99 **Es** 254	100 **Fm** 253	101 **Md** 256	102 **No** 253	103 **Lw** 257